The Ocean's Menagerie

The Ocean's Menagerie

*How Earth's Strangest Creatures
Reshape the Rules of Life*

Drew Harvell

THE BODLEY HEAD
LONDON

1 3 5 7 9 10 8 6 4 2

The Bodley Head, an imprint of Vintage, is part of
the Penguin Random House group of companies

Vintage, Penguin Random House UK, One Embassy Gardens,
8 Viaduct Gardens, London SW11 7BW

penguin.co.uk/vintage
global.penguinrandomhouse.com

First published in Great Britain by The Bodley Head in 2025
First published in the United States of America by Viking in 2025

Copyright © Drew Harvell 2025

The moral right of the author has been asserted

Illustrations © 2023 Andrea Dingeldein
Images on pages 11, 56, and 80 from *Animals
Without Backbones* and *Living Invertebrates* by
Dr. Vicki Pearse. Used with permission of Dr. Vicki
Pearse and the University of Chicago Press.
Insert photographs by Drew Harvell unless otherwise noted.

Designed by Alexis Farabaugh

Penguin Random House values and supports copyright. Copyright fuels creativity, encourages diverse voices, promotes freedom of expression and supports a vibrant culture. Thank you for purchasing an authorised edition of this book and for respecting intellectual property laws by not reproducing, scanning or distributing any part of it by any means without permission. You are supporting authors and enabling Penguin Random House to continue to publish books for everyone. No part of this book may be used or reproduced in any manner for the purpose of training artificial intelligence technologies or systems. In accordance with Article 4(3) of the DSM Directive 2019/790, Penguin Random House expressly reserves this work from the text and data mining exception.

Printed and bound in Great Britain by Clays Ltd, Elcograf S.p.A.

The authorised representative in the EEA is Penguin Random House Ireland,
Morrison Chambers, 32 Nassau Street, Dublin D02 YH68

A CIP catalogue record for this book is available from the British Library

HB ISBN 9781847927712
TPB ISBN 9781847927729

Penguin Random House is committed to a sustainable future
for our business, our readers and our planet. This book is made
from Forest Stewardship Council® certified paper.

I dedicate this book to marine field stations all over the world, but especially Friday Harbor Laboratories (University of Washington) on the Salish Sea, a rich mecca for invertebrate research and study, where I had my beginnings as an invertebrate ecologist and maintain my current research base.

CONTENTS

Preface — ix

1. The Sponge's Pharmacopoeia — 1
2. The Coral's Castle — 25
3. The Sea Fan's Ancient Defenses — 51
4. The Sea Slug's Sting — 77
5. The Giant Clam's Light Trick — 101
6. The Octopus's Shape Shift — 121
7. The Jellyfish's Light Show — 139
8. The Sea Star's Sticky Skin — 167

Epilogue: Spineless Futures in a Warming, Acidic Ocean — 189

Acknowledgments — 211

Notes — 215

Preface

All life began in the sea. Multicelled animals without backbones—the invertebrates—arose in the oceans roughly 700 million years ago, and for upward of 200 million years, they were the only game in town. No animals lived on land, and vertebrates, in the form of fish, had yet to emerge. A menagerie of otherworldly creatures evolved during this very long stretch of time, becoming the ancestors of the marine animals we recognize today as sea anemones, starfish, snails, corals, clams, crabs, and worms. Calling them ancient doesn't even capture the mysterious, deep history of life on our planet. Although emergence of the actual first animal is buried in deep time, all these creatures have prospered for at least 580 million years since their original forms appeared in the Precambrian era, and countless new species have evolved from each group. Plenty of time and circumstance allowed these animals to perfect staggering biological feats unlike any we see on land. There are animals that

PREFACE

photosynthesize like plants, animals that transform wavelengths of light with engineering precision, animals that build castles of glass or limestone using seawater. There are animals that conjure really strange and powerful chemicals that we now use as drugs. I sometimes call the different processes that produce these feats "magic tricks" because, despite our best science efforts, some parts of the essential biological processes are enigmatic, controversial, or shrouded in mystery.

When spineless sea creatures emerged in the ocean over 600 million years ago, their exceptional, multicellular biology upended the balance of power that had previously prevailed beneath the waves. Before them, life in the oceans was dominated for billions of years by single-celled bacteria and protozoans. As the first larger, multicelled animals such as ancestors of jellyfish, sea anemones, and sponges emerged, they transformed the world. Like superheroes, these new animals possessed novel powers that expanded the natural world and allowed them to persist for hundreds of millennia to survive in the oceans today.

In the words of E. O. Wilson, "The love of Nature is a form of religion, and naturalists serve as its clergy. . . . Grant nature eternity on this planet, and we as a species will gain eternity ourselves." As a Cornell University professor teaching invertebrate biodiversity and marine ecology, and as the curator of the Blaschka invertebrate glass collection, I have been privileged to travel the world's oceans on the hunt for some of its most unusual creatures. C. S. Lewis urges would-be writers to tackle what they know and love. Marine invertebrates are my life's passion and somehow, it's the very ancient and seemingly simpler ones that I find most captivating and

PREFACE

surprising: the sponges, corals, jellyfish, and sea slugs. Although less well-known than beloved whales and dolphins, the spineless biota manages the balance of nature in the oceans through creating habitats like coral reefs, controlling the flow of energy through entire food webs, and engineering transformations that baffle scientists to this day.

The ideas animating this book began to drift in my mind many years ago, when I was in college and first fell in love with life in the sea. I became fascinated with the many shapes, body plans, and glittering colors I saw when exploring the rich tide pools of the Pacific coast, and when I did my first scuba dives. There were animals without heads, legs, or eyes. Creatures that looked more like flowers than animals. Animals that had shells and fearsome claws for protection and others that seemed completely vulnerable but were loaded with the most toxic chemicals on the planet. My fascination led me to study marine ecology in graduate school and as a professor at Cornell. The more I learned about the critters in the oceans without backbones—the corals, sponges, worms, jellyfish, clams, crabs, and octopuses that make up 99 percent of diversity in the ocean—the more awe I felt in seeing the marvelous adaptations they had for getting food, reproducing, and avoiding their killers in an environment spinning with an abundance of life. I spent thousands of hours underwater studying how spineless animals run the economy of undersea worlds, and reporting back to an audience of other scientists. A committed reader myself, I longed to share my knowledge and fascination with a broader audience.

In 2014, after *The New York Times* published some of my essays about underwater research in Indonesia, I finally took the first step

PREFACE

into authorhood. My topic sat in plain sight for many years before I saw it. As curator of Cornell's Blaschka Glass Invertebrates, I was in charge of restoring and exhibiting almost six hundred glass figurines depicting particular species of marine invertebrates in fine and accurate detail. The glass masterpieces had been created over many years in the late 1800s by a father and son team, Leopold and Rudolf Blaschka. The story of the Blaschkas and how they created what was essentially a time capsule that we could use to measure changes in the diversity of the tropical and temperate oceans was a quest I described in a book. It was published in 2016 as *A Sea of Glass*.

My next book came out of the terrifying experience of seeing some of my most powerful and iconic invertebrates decimated by a disease epidemic that spread without warning over the West Coast of North America in 2014. As a specialist in marine diseases and leader of a large network of scientists, I was deeply involved in investigating the epidemic and its impacts. Afterward, I developed a book about underwater epidemics and implications for marine life in a warming and increasingly stressed ocean. That book, *Ocean Outbreak*, was published in 2019, just ahead of our own large human pandemic.

Once *Ocean Outbreak* was published, I thought more about an idea simmering in my head from my earliest dives underwater—each kind of marine invertebrate exploded with an extraordinary ability that was somewhere between impossible and magical and made it what it was. My brain ignited with the extent of biological innovation on display in the invertebrates and the sheer wonder of so many fundamentally different body forms built over evolution-

PREFACE

ary time and completely unrepresented on land. I thought about how unusual and shrouded in mystery the core biological adaptations of many invertebrates are, and realized they are more fantastic than the superpowers of the Marvel Comics heroes. This portfolio of strange biologies has given rise to stunning new approaches to engineering and medicine, and the more we look, the more we realize we are only at the beginning of applying new innovations. This book is a journey through my favorite invertebrate body plans in oceans of the world, tapping into the spectacle of life's diversity, the mystery of how these unusual biologies work, and the promise for new innovation.

I start this story as a college student, studying sea slugs in the cold waters of the Salish Sea, at a remote laboratory run by the University of Washington in the San Juan Islands. My big break came as a graduate student, when I was invited on a research expedition to live underwater on a Caribbean reef for a week. As a Cornell professor, I ran expeditions in oceans from Mexico to the Coral Triangle and searched for matches to our Blaschka glass masterpieces. Recently, the enchantment of the Salish Sea and catastrophes of nature brought me back to the same cold, northern waters where I began this journey. The beauties here, and those from all along the way, have not faded.

As an undergraduate student looking for something to do in 1979, I found my people and stumbled on a new life at a place where medicine, biology, and the marine sciences overlap. I entered a hive of research and discovery at a field station where hundreds of undergraduates, PhD students, and researchers from around the world toil on a 24/7 schedule set by the moon's grip on tides and

PREFACE

the circadian rhythms of undersea creatures. Each of the ten research lab buildings is equipped with flowing, bubbling fresh seawater pumped into giant aquariums, housing every ocean creature imaginable, from sea anemones, sea stars, urchins, jellyfish, snails, sharks, and gobies to seagrass, giant kelp, and microscopic algae. In some laboratories, sharks, rays, and ratfish swim in huge round tanks the size of small swimming pools. In other laboratories, giant Pacific octopuses lurk in dark rooms designed as their nocturnal habitat. Other laboratories are free of corroding seawater and outfitted with the highest-resolution imaging and optical microscopes and scanners and DNA sequencers. This is one of the world's notable marine biology field stations, the University of Washington's Friday Harbor Laboratories. During the next forty years, as an undergraduate, graduate student, postdoc, faculty researcher, and educator, I watched scientific discoveries blossom, research teams collaborate, and friendships expand. It's a continuing privilege to live and work in the simultaneously highly regimented and strangely free-form culture of brilliant scientists.

I first came as a college student to work on the ecology of invertebrate larvae. In the beginning, starfish, snails, corals, and barnacles start life at one *fiftieth* the size of a ladybug and are invisible without the aid of a microscope. I had my own beginning studying these tiny babies, called larvae. These larvae leave home on a voyage that can take them thousands of miles as they cross entire oceans and battle innumerable foes. When they find the right spot,

PREFACE

they metamorphose back to a form fitted for a sedentary life at the bottom of the sea, where they will spend the remainder of their time. How do these minuscule larvae navigate this epic ocean quest? How do they know where to settle?

Ironically, I asked these questions when I was making the first big decisions of my own life, studying sea slug larvae as an undergraduate intern at Friday Harbor Laboratories on San Juan Island, nestled in the Salish Sea, north of Seattle. The larvae and I were both so young. How were they to know where in the gigantic ocean they should stop swimming? Once larvae transform to adults, they never travel again. Sometimes I felt like one professional choice might be a pivotal turn on my own journey as well. Still, every voyage must ultimately just begin.

These smallest beginnings of spineless sea creatures rely on the sensory capability to detect the right environment. The sea slug I studied, the frost-spot nudibranch, was an odd little predator that brings a strange talent to the arms race with its prey: it is highly specialized to eat only a single species. It's as if a new human were born who eats only monarch butterflies and can't survive unless they find that one kind. The sea slug's survival relies on the ability of the tiny larva to use its sensors to go all in, to link its entire life with this coral-like animal, called a bryozoan, that lives on kelp blades. Once it finds the lacy white calcified bryozoan, the larva metamorphoses to an adult form of a sea slug and blends in. It becomes completely invisible on the surface of its prey. Hiding in plain sight, it attacks the bryozoan.

But its prey is no sitting duck. My entire PhD thesis was built around the remarkable superpower of this very tricky prey, the marine bryozoan *Membranipora membranacea*, to evade its larval hunter.

The bryozoan is more poetically called the "moss animal." It looks a lot like a coral, with hundreds of tiny polyps sheltered in rigid calcareous boxes; and indeed, also like a coral, its ancestors built entire reefs 400 million years ago, in the Devonian period. Although they can't move, when attacked by a predator, bryozoans have astonishing defenses that reach the limits of our biological expectations. I made my mark as a scientist studying this dramatic showdown between the attacking sea slug and the defending bryozoan. This seemingly simple project set my compass for studying arms races between species and uncovered deeper biological processes of chemically inducible defenses like immune systems that won me a job as a professor at a major university.

Working on the remote, cold shores of the Salish Sea, I cultivated a small farm of bryozoans and exposed half of them to the frost-spot nudibranch. Only three days into the experiment, I noticed something that would take me on a path of discovery—the experimental bryozoans exposed to the predator actually grew a forest of long, dense spines. And they grew more every day. These tiny creatures were shape-shifting before my eyes, changing from what looked like one species to another in the space of days. How did this happen so fast?

This discovery—how the prey bryozoan used chemical detectors to change into a form as different as another species and striking right to the heart of our biological species concept—was the turning point of my career. It inspired me to see nature as full of

PREFACE

mystery and the seemingly impossible. It showed me that experiments could uncover the extraordinary dance of life and the precision of biological interactions cycling beneath the waves. Now I wanted to look everywhere underwater for other discoveries.

Here's why it's worth learning more. Biological tricks and marine superpowers appear to bend expected rules of biology. This gives us a new view of the capabilities and opportunities of life. This possibility makes me stop short with excitement. It's like the feeling described once by Lewis Thomas in *The Medusa and the Snail*: "They remind me of the whole earth at once. I cannot get my mind to stay still and think it through."

If we quiet our minds and stay still, we can start by considering, What are some rules around being an animal? Scientists categorize all the creatures, from sponges to sharks to tigers, as animals. What makes them so?

Historically, scientists have struggled to define what is or is not an animal and to divide these creatures into orderly groups that reflect their evolutionary history and current attributes. Aristotle divided animals into those with blood and those without. Carl Linnaeus created the first hierarchical biological classification for animals in 1758 with his *Systema Naturae*, in which whales were classified as fish and all invertebrates except insects were classified as Vermes, which he described as "animals of slow motion, soft substance, able to increase their bulk and restore parts which have been destroyed, extremely tenacious of life, and the inhabitants of moist places. Many of them are without a distinct head, and most of them without feet. They are principally distinguished by their tentacles (or feelers). By the Ancients they were not improperly

PREFACE

called imperfect animals, as being destitute of ears, nose, head, eyes and legs; and are therefore totally distinct from Insects."

In 1874, Ernst Haeckel gave the invertebrate groups their proper respect and divided the animal kingdom into the multicellular Metazoa (now synonymous with Animalia) and the Protozoa, single-celled organisms no longer considered animals. Many modern animal groups, called phyla, became clearly established in the fossil record during or soon after the Cambrian explosion, which began around 542 million years ago. Animals are defined as multicellular organisms with nucleated cells in the biological kingdom Animalia. (I was curious to see what ChatGPT would say characterizes "animal"; it said animals usually consume organic material, breathe oxygen, and reproduce sexually. I went back to explain to ChatGPT that this was inadequate and that marine invertebrate animals sure do bend these rules!) The biological classification of animals now relies on advanced techniques, such as molecular phylogenetics, which are effective at revealing the ancient relationships and evolution within and between the thirty-five dominant animal groups.

Scientists divide animals into those with a backbone (the vertebrates) and those that are spineless (the invertebrates). The ones most people see regularly on land and know best are the vertebrates. They are a very cohesive group that includes fish, sharks, lizards, birds, and mammals, with quite a narrow range of biological capabilities—one small group among thirty-four others. They are an evolutionarily new group, most diversified on land, and radiating from a fish ancestor that came ashore. But the vertebrates are essentially all variations on the same theme: a head with a brain and often eyes, a backbone, and often four limbs in equal numbers

PREFACE

on both sides of the body. It's the invertebrates, in the thirty-four other groups, that are bursting with ancient and unusual capabilities, born of hundreds of millions of years of living in the sea. And it is in considering these capabilities that we slam into the very nature of what it means to be an animal, because some of these do not consume organic matter and instead derive their nutrition from the sun like a plant; many do not breathe oxygen; and many reproduce asexually, instead of or in addition to sexually.

One way invertebrates transcend the normal rules of animal biology is through partnerships with entities across the kingdoms of biology. The best known of these are the symbioses with bacteria and microalgae. Stunning superpowers born of this partnership are the bioluminescence of squid and ostracods, the potent cancer drugs found in sponges, and the photosynthetic and reef-building capabilities of corals. We now have a relatively new word in biology to account for and refer to these networks of powerful relationships, defining new ways of thinking about what it means to be a single animal at all. *Holobiont* is the word for the entire sphere of life that encompasses an organism. The coral holobiont includes the actual coral polyp, plus its photosynthesizing algae, plus the myriad bacteria whose actual roles are not yet known, plus the possibility that there are also symbiotic viruses on those bacteria or algae. The complex network of holobiont interactions is an unknown frontier in biology.

Along the way, both the sea and I have changed. Some stars of our oceans have fallen and lost their place on this planet. The intensifying pace of climate change and human interference in our oceans have changed our lives on land. My colleagues and I have

PREFACE

transformed from adventurers in pristine places to conservationists in service to preserving the waters in which life can thrive and evolve. Join me on a deep dive to explore the ancient biological powers of spineless animals, from sponges, corals, and jellyfish to octopuses and sea stars.

The Ocean's Menagerie

Marine sponges come in many shapes and sizes, including vase sponges, barrel sponges, those encrusting sponges flat across the bottom, and those overgrowing corals.

1.

The Sponge's Pharmacopoeia

On a hot Bahamas day in 1987, at a site called Chub Cay, we hauled our scuba tanks off the 155-foot-long research vessel *Columbus Iselin* and into a small rubber Zodiac. We motored to nearby rock islands, anchored, and donned masks, fins, snorkels, and tanks. The four of us then dropped backward off the Zodiac into the shallow water. We divided into two teams and swam in opposite directions. The sun was out, the water was warm and clear, and it seemed I could see for miles across the shallow reefs. The reefs were healthy and populated with big fish, little fish, fast fish, and all colors of darting, dashing, diving fish. They were a delight but also a distraction from what we were looking for: sponges containing chemicals with potential for use as human drugs.

I swam slowly above a bottom carpeted with brown, green, and yellow hard corals, purple and tan soft corals, red and brown encrusting sponges, and purple vase sponges. The heads of hard corals towered around me, dotted with colorful sponges and sea squirts and wreathed with red and green algae. A little deeper, we started to see huge barrel sponges, some as tall as a human, looming large in the distance like sentinels of the reef. I was satisfied to see that sponges were on full display in a variety of forms.

The dominant sponges, brown in color, encrusted the bottom, carpeted rocks, and coated all sides of some coral heads. Some of the red and yellow sponges were ropy and crept in narrow tendrils along the bottom or upright as swaying columns. I stopped to inspect a particularly beautiful purple vase sponge in the genus *Verongia*. It was over a foot high, smooth skinned, pale lavender, and perfectly vase shaped. I hung suspended in mid-water, feeling like I had traveled back in time to an era more than 500 hundred million years ago, at the dawn of animal evolution, when sponges ruled the seas.

The evolutionary leap from single-celled to multicellular organisms is one of life's most consequential innovations. It may have occurred more than once, each event forming a distinct lineage. The earliest forms of the animal lineage may not have been classifiable as "sponges," but among all the animal sub-lineages alive today, that of the sponges and ctenophores are considered the most ancient. The first spongelike animals in Precambrian seas were probably just tiny aggregations of cells with only modest specialization of function. A fossil recently found in China and dated to 600 million years ago appears to be the trace of such an organism. Not until the Cambrian era, at 520 million years ago, did sizable sponges appear in the fossil record. An ancient sponge lineage, called Archaeocyatha, dominated the seas in the Lower Cambrian for about 20 million years and diversified into hundreds of species of double-walled cup and vaselike shapes. With skeletons made of calcite, they created the planet's first reefs. Archaeocyathid reefs would have supported evolution and invertebrate diversification by creating a complex habitat. The archaeocyathids dwindled in the later Cambrian but gave way to an explosion of sponge forms, many appearing remarkably like those that exist today.

The archaeocyathids may have been the first invertebrates that may have forged partnerships with microorganisms. Living in shallow, sunlit waters, they may have hosted photosynthesizing cyanobacteria for a share of the sugars their guests produced. So this most ancient of our invertebrate lineages got its start in a symbiotic partnership with bacteria. It would prove to be a model emulated many times throughout the history of animal evolution.

The lack of mobility made early sponges sitting ducks for predators and competitors; evolving effective chemical defenses gave these sponges an important adaptive advantage. Pressure from predators is one possible reason why modern-day sponges are masters of chemical innovation. Another explanation may be their long tenure in an ocean filled with pathogenic microorganisms and a body design that leaves all cells exposed to those pathogens. Making toxic chemicals would have been a great advantage for a sessile creature to fend off biological threats like competitors, predators, and pathogens.

While I have been discussing sponges as if they are all the same, modern sponges are a diverse group, with over 5,000 species described worldwide. There are more than 230 sponge species in the Caribbean and 830 in Indo-Pacific Indonesian waters.

As I studied the purple vase sponge, I thought about the huge volume of water it draws through its pores to filter for food and expel through the central opening. You can't easily see the currents the sponge creates unless there are a lot of particles in the water. My dive buddy was a student intern from Cornell. I nudged her with my elbow and nodded—a sort of "watch this" gesture. I pulled a syringe loaded with a fluorescent green dye from my dive vest and ejected a small blob of it near the base of the sponge. The blob was quickly sucked in through the side of the sponge. Through her mask I could see her eyes widen. I put up a finger in a gesture of "wait for it." After about forty seconds, a single, continuous stream of fluorescent green dye streamed out of the top opening of the sponge. I handed her the syringe to try for herself, and we spent another five minutes transfixed by the currents rushing through the sponge. Then she realized there were sponges around us in other shapes and sizes, and she started experimenting with the dye to see how the flow rates and patterns varied. Mission accomplished! I had wanted to share the sheer delight of watching the pumping action of sponges. The other divers were leaving us behind, so we had to stop messing with the dye streams and start our collecting work. While I have never liked breaking off parts of sponges, they regenerate rapidly, so any small piece we took would regrow.

The medium-sized sponges with multiple upright columns like champagne flutes were easy to collect by breaking off a small column at its base. For the larger barrel sponges, we used our dive knives to pry off an edge. The encrusting sponges were the hardest to sample. They live tightly attached to the reef and often have dense bone-like spicules embedded in their tissues.

Many animals, like tropical frogs and caterpillars, have bright colors and contrasting patterns that advertise the presence of chemical defenses in their bodies. If potential predators have learned to associate the colors with bad taste or sickness, they are less likely to attack. We suspected that bright colors in tropical sponges were a similar adaptation. They could signal to a hungry fish that the sponge it is considering for lunch packs a harmful punch. It turns out that the kinds of chemicals that are toxic also have associated properties on biological systems that could make a compound useful to medical science, so sponges with strange and dazzling colors were our main targets.

Some of the brighter sponges were small and lived in more hidden places, like the crevices and cavities under coral heads and the darker hollows under large overhangs. We had to reach into these recesses to lever off the sponges with a knife. The gloves we wore protected our hands from the chemicals and sharp spicules in the sponges—and also from the moray eels that often lurked in these same spots. As we collected each sponge, we jotted notes on our underwater slates about the depth, location, and appearance of the sponge. Each sample bag was numbered to relate to the notes and placed in a mesh collecting bag. We continued swimming along the edge of a shallow reef at about forty feet. After fifteen minutes, we noticed we were being followed.

A pair of gray angelfish swam close to see what we were doing. There was no missing them, because they are almost a foot tall and were right beside us, like nosy shopkeepers watching for shoplifters. Gray angelfish are gray with dark spots on each scale, dark eyes set in a pale gray face, and long trailing dorsal and anal fins trimmed in blue. It's usual to see them in pairs, because they mate for life. Perhaps they hoped we would dislodge a rock and expose a sponge for them to eat. They darted beside us around the shallow reef, picking constantly with their tiny protruding mouths at things we could not see. Knowing that sponges are their primary food, we decided to turn the tables and follow the angelfish to see where they would lead. They moved slowly along the reef, occasionally taking a couple of bites of the most common sponges, the ones we could spot easily ourselves. Then our gambit paid off when one of the fish darted under a coral head. I took a look and spied a small, bright yellow sponge we hadn't seen before.

Compared with the typical research dive, the kind of underwater collecting we were doing was easy and very enjoyable, like a treasure hunt. Usually in undersea research work, we have a fixed goal and extensive data sheets to fill in. It requires laser focus and careful timing to capture all the data needed from one or several survey transects. With a set amount of air in scuba tanks, bottom time rarely exceeds an hour. There is almost never enough time to accomplish the research goals, so you can't really look around and appreciate the habitat and all the organisms you aren't studying. My diving partners often saw sharks, manta rays, turtles, and other exciting things, but I always missed them, my eyes trained on either the bottom or the data sheets on my slate. So exploratory dives like this where I could trail gray angelfish and explore the richest habitat on our planet were pure joy. I was in my happy place, looking for interesting new sponges and pondering the diversity of sponge form, color, and evolution. As a young professor at Cornell University, a year into the job, I was reminded that doing science could be adventure at its best.

By the end of our dive, we had each collected fragments of about twelve different sponge species. As an ecologist and conservationist, I hated prying these sponges from the reefs they were thriving on. I was mindful that if we struck it rich and found a species with important activity against cancer, that species would be ripe for overcollecting. But as a conservationist, I was aware that finding this kind of value from the ocean is key to fostering a preservation ethic among members of the public. I also knew that drug discovery was not a major threat to biodiversity, because drug companies would rather not rely on uncertain, rare, natural sources and eventually synthesize useful chemicals in the lab.

We circled back underwater to our anchored Zodiac, climbed in, and motored back to the laboratory ship. When we bumped against the ship and handed up our empty scuba tanks, gear, and buckets with specimens, the lab chemists, student assistants, and drug assayers came out from the lab to see our underwater haul. They crowded around the buckets to see the treasure trove of red, maroon, orange, yellow, blue, brown, and purple sponges. We each showed a few of our most exciting finds and answered the barrage of questions from the chemists about where we had found them and how we had spotted them underwater. We were pleased that they seemed to appreciate the natural history expertise it took to find new, unusual, possibly rare sponge species. The scientific co-learning was invigorating on both sides and created a culture of exploration on our research ship, one that would be continued in conversation over shared meals and beer. For now, however, everyone needed to move quickly because some of the active chemicals in our samples might degrade in full sunlight or air. We hauled our buckets from the outer deck into the air-conditioned lab, which was loaded with equipment and instruments.

We handed our sponges over to the chemists, who briefly air-dried the samples to remove the extra water that would interfere with the extracting chemicals and then extracted the sponge compounds with solvents like ethanol, methanol, and dichloromethane, each able to pull out slightly different chemicals. These chemical cocktails were then passed along the same day to the drug assay folks to test. Because sponge chemicals that disrupted cancers had been previously discovered, a prominent assay on this trip was for cancer drugs. In cancer, cells divide uncontrollably, forming masses like tumors. So the goal in searching for cancer drugs is to find chemicals that slow or prevent cells from dividing. Our research expedition was focused on finding chemicals with activity against cancer, and so scientists hunting for cancer drugs brought along petri dishes loaded with rapidly dividing cancer cell types. The assay involved pipetting small amounts of sponge extracts into the dishes and seeing which ones slowed or stopped the cells from dividing.

The next day there was a buzz of excitement. Out of the roughly twenty-eight sponges being tested, extracts from three were slowing the cancer cells, and the chemists wanted us to collect more of these target species. I wish I could report that we had found an actual anticancer drug, but the chemicals from these sponges were not just slowing the growth of the cancer cells, they were killing all the cells in their path. In the end, these chemicals were too toxic to be good drug candidates.

Humans have long sought out organisms in nature as sources of medicine. We have been masters at discovering medicinal value in leaves, roots, fruits, mushrooms, animal livers, and other organs and using these materials to treat fever, colds, congestion, infections, and other maladies. Aspirin, for example, is derived from a chemical in willow bark, and morphine from poppies. Traditional cultures' natural pharmacopoeias are sourced mainly from plants, fungi, and terrestrial animals—the organisms with which they have closest contact. Only a few traditional medicines come from the sea. But over the last several decades, scientists have found that some groups of marine organisms may be the richest sources of drugs on the planet.

Of all the animals that live in the oceans—crustaceans, cnidarians, echinoderms, mollusks, worms, tunicates, fish, and others—none seem to offer as much pharmacological potential as sponges. Our knowledge of the curious superpower of sponges as master chemists has a fascinating history with surprising twists of scientific discovery.

In the 1980s, scientists at the National Cancer Institute were noticing that chemicals from marine invertebrates were so different and biologically potent that they were becoming a leading source of new ideas and leads for designing new drugs. At the time, the highest "hit rate" for new anticancer drugs came from tropical sponges. Hence our research expedition was partly funded by the National Cancer Institute (with Professor William Fenical of the Scripps Institution of Oceanography as principal investigator) and focused on the potential of finding new chemicals from sponges with cancer-disrupting activity. The natural places to look for such sponges are coral reefs because they are the most diverse ecosystem on our planet and typically support many different sponge species.

I was along as an ecologist on this drug-discovery expedition to help find and identify useful coral and sponge prospects to test. My personal research goal wasn't so much looking for drugs as it was understanding how these animals used their unusual chemicals in nature. On every dive, my mind was focused on understanding how corals and sponges used potent chemicals to fight for space on the reef and as defenses against the predators that would eat them and the diseases that would infect them.

What are these creatures called sponges? It's hard to fathom that the integrated collection of cells that constitutes a sponge is an animal. A sponge lacks eyes, limbs, and a head, and it can't move. It has no real organs. It is, in the simplest terms, a collection of flagellated cells organized around water flow channels. The flagella are the tiny hairs that beat in a coordinated way to move the water inside the sponge. They are myriad mighty pumps. Water moves in one direction, in through the small side pores, to chambers of flagellated cells, and out the top of the central chamber. The incoming current contains bacterium-sized food that is caught and collected on collars of the flagellated cells. Although sponges are the simplest multicelled life-forms, they carry on biologically sophisticated life processes.

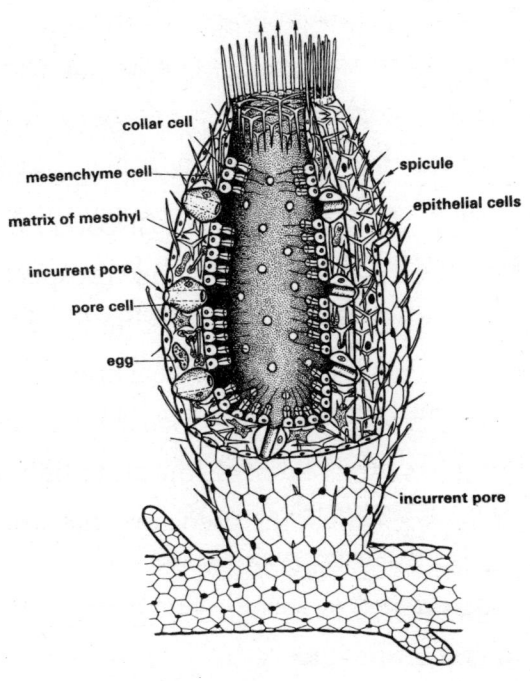

Diagrammatic, simplified view of the inner workings of a sponge, showing the incurrent pores, the excurrent vents, the collar cells that power the pump, and the spicules that form skeletal elements.

DIAGRAM COURTESY OF VICKI PEARSE, *LIVING INVERTEBRATES* (PALO ALTO, CA: BLACKWELL SCIENTIFIC PUBLICATIONS, 1987)

What sponges lack in organ systems they make up for in their extraordinary multipurpose cells and the elegance of their water flow systems. These systems are why it is so breathtaking to use dye streams to visualize the rate at which water is pumped through a sponge. While sponges are essentially just collections of cells, each of those cells is totipotent; like the Marvel superhero Deadpool, who has superhuman strength and a regenerative capability to regrow parts as long as even a single cell in his body survives, each sponge cell can develop into multiple forms, take on a variety of tasks, and switch functions with a change in current speed. Over the course of its life, a single sponge cell can switch from forms that engulf food to those that reproduce or make a skeleton. At the food-collecting sites, each cell collects its own food but can then change into another cell type and carry food around inside the sponge's body. These multipurpose cells can also detect pathogenic agents, engulf and destroy them, or launch a chemical counterattack with their primitive immune system. The cells can also guide the water currents within a sponge. By comparison, human cells are totipotent only in their early embryonic phase, when a stem cell can regenerate an entire human and all its organs. As our cells age, they become programmed to specific functions and lose the power of regeneration and totipotency.

Inside a sponge, the synchronous beat of the many flagellae creates high-speed currents that are smoothly directed through increasingly lower-flow channels to reach the almost still, low-speed sites of food collection. Engineers studying the design of pipes in the sponge water system marvel at the impressive design. The low flow speeds accomplished at transfer sites result from the greatly increased cross-sectional area and thus greater volume of these small channels, just like the high surface area of capillaries slows down blood flow at oxygen-exchange sites in a human circulatory system. On the way out of the sponge, in contrast, the excurrent waters jet at high speed through the much-reduced cross-sectional area of the central column. These passively changing water speeds fit the functional needs of the sponge perfectly. While all this is going on inside the sponge, it never moves from its anchored spot. Sponges are all sessile, sitting seemingly motionless on the bottom of our oceans, lakes, and streams.

The chemical novelty of marine invertebrates is practically limitless, and probing the scope of invertebrate chemistry is still a scientific frontier. In no other group is the portfolio of valuable chemicals more developed than in sponges. That is why I call the capability to produce potent biologically active chemicals a sponge superpower. Scientists label some of these biologically active and often toxic chemicals "secondary chemicals" because they have no central metabolic functions for the sponge; they don't help it breathe or eat or reproduce. Despite being secondary in this sense, the functions of these chemicals are often of great importance to sponges—reducing inflammation, slowing cell division, deterring predation, and reducing sun damage. These are complex chemicals, with backbones made of more than twenty carbon atoms, elaborate three-dimensional shapes, and atoms of special elements like bromine or chlorine often attached. Making, handling, storing, and deploying such potent chemicals is like handling a bomb or dangerous toxin and requires biological processes and containment that are not well understood; I call them biological tricks.

As a research scientist, I learn over and over that we do not have all the answers and there are still big mysteries of life on earth to solve. I watched over the course of my career as the mystery of how these chemicals are produced took surprising turns. When we were collecting sponges in the Bahamas in the late 1980s, we assumed that any promising chemicals we found in the sponges were made by the sponges themselves. About a decade later, scientists determined that many of the bioactive chemicals found in sponges were made not by the sponges but by bacteria living inside them. In these cases, the trick was on us; the tiny bacteria hidden in the sponges were the real chemists. It was as if, long ago, the sponges had made a deal with bacteria: make us defensive chemicals and we will provide you with food and shelter. Similar discoveries in nearly every branch on the tree of life are being made all the time: we are learning that many plants and animals depend in various ways on algae, bacteria, and fungi that live inside, on, or around their bodies.

I think I am correct that the first work to attribute the production of these useful chemicals to the bacteria instead of the sponge was by John Faulkner's group at Scripps. I met John on some of our earlier research cruises to the Bahamas. His group investigated the localization of natural products within sponge-to-microorganism associations by isolating and chemically analyzing cell populations within sponge samples. Through this approach, the researchers located the cytotoxic chemical swinholide A and the peptide theopalauamide inside two bacterial strains. Both bacterial strains were isolated from the sponge *Theonella swinhoei*. Their work launched new directions in research.

New studies continue to reveal cases where microbes are the

chemists making the clinically significant bioactive chemicals found in marine sponges. These reports come from tropical oceans around the world, including the Caribbean, the Mediterranean Sea, the Great Barrier Reef of Australia, the South China Sea, Indonesia, Papua New Guinea, and the Indo-Pacific region. These important studies are filling in key knowledge gaps about the biology of novel chemical synthesis in sponges and the roles of the different players.

The bacteria and microalgae living in sponges are turning out to be a key part of sponge biology, with roles that go beyond the manufacture of secondary chemicals. Like corals, some sponge species harbor algal symbionts that photosynthesize, as well as a multitude of diverse bacteria. More than 50 percent of the total weight of a sponge can be bacteria. Sponges offer protection and nourishment to their microbial inhabitants, and some of the microorganisms return the favor by performing jobs such as fixing nitrogen or photosynthesizing to make food for the sponge. In many cases, it is the bacterial symbionts that synthesize the secondary compounds that protect the sponge host against predators and infection. While scientists have long known that bacteria create biologically active chemicals, in the last two decades, scientists have learned new aspects of bacteria's dazzling ability to craft elaborate and highly bioactive chemicals with precise effects on the biology of other organisms, such as antibiotics that slow or kill bacteria and anti-inflammatories that reduce swelling.

Finding novel drugs made by microorganisms is like a lottery. For land-based research, bacteria in the genus *Streptomyces* are a hot ticket. They produce a wide range of secondary chemicals with antibiotic, anticancer, antifungal, and other properties. There are also diverse strains of *Streptomyces* in marine sponges. Given the treasure trove of drugs that have come from terrestrial *Streptomyces*, the marine ones may offer huge potential. Indeed, endosymbiotic *Streptomyces* bacteria have proved to be one of the richest sources of bioactive chemicals in marine sponges.

In one study, ninety-four strains of *Streptomyces* were found in just four marine sponges (*Callyspongia diffusa*, *Mycale mytilorum*, *Tedania anhelans*, and *Dysidea fragilis*). Of these ninety-four, fifty-eight could disrupt bacterial growth, thirty-six disrupted fungi, and twenty-seven did both. Of the fifty-eight strains that disrupted bacteria, thirty-seven inhibited *Bacillus subtilis*, forty-three inhibited *Staphylococcus aureus*, ten inhibited *Vibrio cholerae*, and ten inhibited *Escherichia coli*. In the race to find new drugs to combat increasing antibiotic resistance in disease-causing bacteria, species of *Streptomyces* have been an extraordinary source of new leads. *Streptomyces* belongs to the phylum Actinomycetes and produces 80 percent of natural products derived from Actinomycetes, itself estimated to produce "about 70% of the naturally derived antibiotics that are currently in clinical use." Even one of our common sponges in the Pacific Northwest hosts a drug-producing *Streptomyces*.

Halichondria panicea is a bright green intertidal sponge I see every time I go to the shoreline at low tide in the Pacific Northwest. Scientists discovered that *Streptomyces* strain HB202 from *H. panicea* produces a cytotoxic chemical called mayamycin. Mayamycin was noticed because it stopped clinically relevant bacterial strains from growing, some of which were resistant to multiple existing antibiotics. Further tests showed impressive and diverse anticancer activity. Mayamycin disrupted cell division in eight different human cancer cell lines, including hepatocellular carcinoma, colon adenocarcinoma, gastric cancer, non-small cell lung cancer, mammary cancer, melanoma, pancreatic cancer, and renal cancer. Mitomycin, a synthetic derivative of mayamycin, is currently a medication used in the treatment of anal, bladder, and breast carcinomas; head and neck malignancies; and some other gastrointestinal carcinomas.

Although sponges and other sessile marine invertebrates and their endosymbiotic bacteria produce many compounds with the potential to become anticancer treatments—so-called drug leads—it is a very long road to market. In 1986, scientists isolated a compound they called halichondrin B from the sponge *Halichondria okadai*. Halichondrin B had potent antitumor activity, inhibiting the growth of tumor cells by interfering with the tiny filaments that control cell division. However, clinical trials of halichondrin B were not initially possible, as harvesting or aquaculture of *Halichondria okadai* did not yield great enough quantities of the compound, and the chemical structure was delicate and easily disrupted. In 1992, researchers synthesized halichondrin B and then later, in 2004, developed a simplified and "pharmaceutically optimized" version of the compound they called eribulin mesylate, which was put into clinical trials against specific types of breast cancers. Finally, twenty-four years after halichondrin B was discovered, eribulin mesylate (under the brand name Halaven) was approved by the FDA in 2010 for patients with some kinds of metastatic breast cancer.

Much as they would like to, chemists cannot just make up chemical structures that might have useful biological activity. The relationship between chemical structure and activity in biological systems is just too complex. They need to first find the compound in nature; then—in a process that is often very difficult and time-consuming—they can produce it synthetically and tinker with its structure in various ways.

Another example of an anticancer compound from a common marine sponge is dysideanone D. It comes from *Dysidea avara*, a large tropical Pacific sponge with foot-high columns projecting from its base. Dysideanone D inhibits growth of several kinds of tumor cells. It is better at inhibiting cervical carcinoma cells than present drugs and is being considered for treating cervical, liver, lung, and colon cancers. Dysideanone D is found only in this particular species of sponge, so initially it was thought the sponge itself made the chemical. Then detailed work revealed that it was made by a marine bacterium living inside the sponge. As is the case with many other potent chemicals produced from sponge-microbe symbioses, no one knows what dysideanone D does for either the bacterium or the sponge. Moreover, the terms of the symbiotic partnership between *D. avara* and the bacterium remain mysterious. Does the sponge acquire its bacteria, or are they passed on from its parent as part of reproduction? Does the sponge control the growth and reproduction of the bacterium? Does the sponge control the production of dysideanone D? Is the bacterium responsive to messages from the sponge? Why does the bacterium live only in this one sponge species? As these unanswered questions indicate, endosymbiosis is a research frontier studded with large gaps in knowledge. In my mind, working out the details of the dance of the symbiosis—the costs and benefits of the partnership—is the million-dollar question behind the superpower of making novel drugs.

Many invertebrates besides sponges partner with bacteria, algae, or fungi for vital functions, and these relationships may give us clues to what is going on in sponges. Several animals, such as the shrimp *Palaemon macrodactylus*, actively recruit the bacterium *Alteromonas* as protection against a pathogenic fungus. Similarly, many marine invertebrates rely on the presence of certain bacterial species to induce metamorphosis from larva to juvenile. Without the cues provided by the bacteria, they never reach their adult form. Multiple bacteria have evolved a tight relationship with fruit flies. They live in the fruit fly gut and help digest unwieldy plant material. Many of the genes they inherited from their free-living relatives no longer function, and they cannot survive outside a fruit fly gut. At this point, we say that the relationship has become obligate for the bacterium. We know less about the complex coevolution between sponges and their bacteria but expect similar kinds of coadaptation, including obligate tight relationships for one or both partners. But expectations are different from confirmed relationships: for most endosymbioses between sponges and bacteria, we don't know whether the relationship is obligate or a casual association between tolerant partners.

Understanding how associations among bacteria and animals evolved may reveal unusual biological rules that govern ongoing interactions. We do know that animal and bacterial genomes are intertwined; indeed, many animal genes originated in bacteria and were taken over by their animal hosts, even in humans. An estimated 37 percent of human genes originated in bacteria. Co-opting some of the vast genetic information present in a bacterium affords its host, from sponge to human, a way to readily expand its metabolic functions. Because sponges may have been cultivating partnerships with bacteria, single-celled algae, and viruses for longer than any other invertebrate group, they may also have been the first to acquire parts of bacterial genomes.

Similarly, the great age of the symbiosis between sponges and microorganisms—first seen perhaps 520 million years ago in the presumed relationship between proto-sponge archaeocyathids and cyanobacteria—allows for unusual capabilities to have evolved. Is an extremely ancient relationship with bacteria why sponges stand out among invertebrates for making the most novel and biologically active chemicals? And to focus this question back in time half a billion years: Did sponges borrow the services of bacteria to create the first chemical defense systems, the first functional immune systems?

The key ability in immune system function is recognizing the difference between self and nonself—the cells that belong to you versus the ones invading from the outside. Or at least that's the way the issue has been posed since immune systems began to be understood in the late 1800s. If the origin of the sponge immune system is tied up in an ancient form of endosymbiosis, it puts a very different spin on that issue. Foreign cells that were not at all like sponge cells were allowed—even welcomed—to live inside sponge bodies, and they seemed to have received their entry pass precisely because of their ability to make compounds that could keep out other single-celled invaders that could cause harm to the sponge. Thus, half a billion years ago, before immune systems as we know them, the key issue for proto-sponges was not self versus nonself but beneficial versus harmful. How were proto-sponges able to make that determination? We have no idea. But as we learn more about the importance of microbes in animal-life processes, the answer may take on more significance than we expect.

The chemicals discovered in sponges and brought to market as drugs show the enormous benefit of turning to nature for innovative ideas. Because oceans represent about 90 percent of the habitable volume on our planet and contain an estimated 15 percent of the world's animal biodiversity, marine habitats and organisms are clearly an important source not just of drugs but also of biological insights and genetic resources. The widely recognized value of marine natural products was formalized in the recently negotiated global High Seas Treaty with eighty countries, called Biodiversity beyond National Jurisdiction. This treaty will be essential in designating marine protected areas, establishing a process of environmental impact analysis with the power to stop deep mineral mining and overfishing, and creating a framework for equitable sharing of the benefits that come from future discoveries of useful genetic resources. The high seas remain a relatively unexplored frontier, so there is great potential for discovering valuable resources lurking in the depths.

Working out the section of the treaty pertaining to the fair sharing of valuable future discoveries was the most vexing and challenging part of the twenty-year treaty negotiation. The difficulties lie in understanding and describing the potential of an unknown portfolio of ocean riches. This has to take into account our fast-changing technical understanding of the value of the genes controlling their valuable chemical products and the organisms in which they occur, and then determine how to decide in what political jurisdiction an organism might have originated. These administrative and legal issues are layered on top of the enigmatic biological issues that I have described in this chapter. These include whether a powerful chemical or other product is made by the animal or one of its endosymbionts and, if the latter, how the animal may facilitate the synthesis of the chemical of interest.

We are learning that partnerships with bacteria abound across the tree of life. As the first metazoans to innovate a working partnership with bacteria, sponges have a great deal to tell us as we seek to understand the dance between host and symbiont. It is one of the great challenges in biology, an exciting frontier in the management of natural ecosystems and human health.

Reef-building corals come in a dazzling diversity of colony forms, from encrusting on the bottom to delicate branching, plating, mounding, and towering edifices.

2.

The Coral's Castle

The richest habitats on earth, coral reefs house a quarter of all marine animal species. A coral reef is the place to see a representative slice of the biodiversity that two billion years of evolution have produced—from sliding snails, wiggling worms, clambering crabs, and sedentary sea stars, to flashing fish. The sheer diversity of life on display is a big part of the reason coral reefs have been my favorite places to dive and a top spot for research. Coral reefs are also indispensable resources for humans wherever they occur. They support artisanal fisheries and protect coastal communities from damaging waves and subsequent erosion. As places to swim, snorkel, dive, fish, and watch wildlife, they are the centerpieces of tourism industries providing revenue for tropical countries around the world.

These massive structures so important to human economies and ocean life are built by tiny, soft-bodied animals related to sea anemones and jellyfish. Each individual animal, called a polyp, is smaller than a caterpillar. What coral polyps lack in size they make up for in numbers and physiological ability. Reefs on the scale of human cities exist today because ancestral coral polyps evolved three related abilities that together constitute a powerful superpower. Each coral polyp partners with photosynthetic algae in its skin to harness energy from the sun; that energy then precipitates calcium carbonate from seawater and fashions it into rocklike housing units. As this goes on, the polyps clone themselves, growing large colonies that multiply their efforts many times over. As Charles Darwin wrote in *The Voyage of the* Beagle, a coral reef represents "the accumulated labor of myriads of architects at work."

Just as we saw with the chemistry skills of sponges, the crux of the coral superpower is a symbiotic relationship with single-celled microorganisms that live inside the animals. In the case of reef-building coral, the symbionts are golden-brown algae, or dinoflagellates, called zooxanthellae. The tiny round zooxanthellae, each about twelve microns wide, are packed in lines, studding the skin of the corals. The zooxanthellae use several different pigment molecules, including chlorophyll, to make sugars from sunlight. They provide the coral these sugars as energy for growth and calcification. During the day, sunlight penetrates the shallow water in which corals live, and the sugars made by the zooxanthellae provide the polyps with most of the energy they need for growth and reproduction.

Coral's symbiosis with zooxanthellae likely originated about 400 million years ago as an adaptation for nutrition. Corals live in clear water with low nutrients, and the zooxanthellae provided them with extra food in an environment in which the supply might be uncertain. In this respect, early corals were much like other invertebrates—such as giant clams (see chapter 5)—taking advantage of a partnership with photosynthetic microorganisms. Over evolutionary time, however, corals took their symbiosis-based superpower to new heights. The churning, pounding waves that swept through their shallow-water habitats drove corals to evolve into some of the world's best structural engineers. They started life with small skeletons over 500 million years ago but, with the aid of their symbionts, grew into builders of enormous castles able to withstand extreme wave shocks driven by hurricane-force winds. A single coral colony can stand over forty feet high, and a reef built from countless colonies can extend for hundreds of miles.

I was just on the island of Viti Levu in Fiji, watching huge waves crash on the fringing coral reef, not far from shore. Despite the big waves on the fore reef, I was able to snorkel safely in calm water behind the wave barrier of this gigantic reef. I felt firsthand the security of that enormous, live coral barrier stretching all around Fiji's northwest coastline. The Great Sea Reef is the third-largest continuous barrier reef in the world and stretches over two hundred kilometers. As a natural and organically growing live barrier of the main island in Fiji, it is referred to as the "Protector of Fiji" by many Fijian communities. Not only does it protect the shoreline, but it also contributes yearly between US $6 million and $8 million of fisheries' value. At least a tenth of the entire population directly relies on it for their livelihoods.

As he explored Keeling Atoll in Australia during the *Beagle*'s voyage in 1832, Charles Darwin also expressed an awareness of coral's engineering prowess:

> *Millepora complanate*, the fire coral, grows in thick vertical plates, intersecting each other at various angles, and forms an exceedingly strong honeycombed mass. . . . Between these plates and in the protected crevices of the reef, a multitude of branching zoophytes and other productions flourish, but the *Porites* and *Millepora* alone seem able to resist the fury of the breakers. . . . It is impossible to behold these waves without feeling a conviction that an island, though built of the hardest rock, let it be porphyry, granite, or quartz, would ultimately yield and be demolished by such an irresistible power. Yet these low, insignificant coral-islets stand and are victorious: for here another power, as an antagonist, takes part in the contest. The organic forces separate the atoms of carbonate of lime, one by one, from the foaming breakers, and unite them into a symmetrical structure.

After returning home, Darwin collected his observations in the book *The Structure and Distribution of Coral Reefs*, published in 1842. I feel a strong kinship with Charles Darwin when he pens his awe at the power of tiny animals to create the substance from which whole islands are formed:

> It is not a wonder which at first strikes the eye of the body, but rather after reflection, the eye of reason. We feel surprised when travellers relate accounts of the vast piles & ex-

tent of some ancient ruins; but how insignificant are the greatest of these, when compared to the matter here accumulated by various small animals. Throughout the whole group of islands, every single atom, even from the most minute particle to large fragments of rocks, bear the stamp of once having been subjected to the power of organic arrangement.

Even in 1836, as he marveled at the reefs in Australia, Darwin understood that corals follow precise blueprints to build their skeletons. If you look closely at the coral skeleton under the polyps' colorful skin, you see a highly branched, bright white crystal lattice. The lattice shape is species specific, but all shapes are very strong. On wave-swept shores, a coral reef can absorb 97 percent of the wave energy without breaking. This explains how some of the biggest North Shore surf breaks in Hawaii crash on vast, delicate reefs without damaging them. While we don't have data from Hawaii, a study of the strength of Caribbean corals shows that their fracture strength (the stress at which failure begins to occur) is 1,740 to 12,000 pounds per square inch, with samples from branched colonies being stronger than those from massive, mounding, boulder-shaped colonies. These values are higher than human-produced carbonate construction materials like concrete and synthetic limestone.

Corals don't simply patch together calcium carbonate; each species precipitates the mineral crystals in its own pattern to form elaborate microstructures as complex as lacework and as strong as steel. The precipitation process itself requires energy. Although seawater contains negatively charged carbonate ions and positively charged calcium ions in solution, these two components of calcium carbonate don't join together spontaneously unless the pH is high enough to make the process energetically favorable. To create this favorable environment, tiny coral polyps pump hydrogen ions (which always exist in any sample of water) out of the "working space" between special cells and the surface of their existing skeletons. This raises the pH of the space, allowing more carbonate and calcium ions to bond together to form crystals of aragonite, a form of pure calcium carbonate.

We know sunlight has a role in building the skeleton because experiments show that corals make skeletons faster in the light than in the dark. The sunlight, which drives photosynthesis inside the symbiotic zooxanthellae, also accelerates calcification because photosynthesis absorbs acidic carbon dioxide and thus helps bring the pH to the right levels for calcification. The zooxanthellae, therefore, are essentially solar reactors doing double duty: in the process of making sugars through photosynthesis, they are not only feeding the coral but also speeding up calcification.

The ability to build carbonate skeletons from seawater is shared across many branches of the tree of life. Protozoans build tiny water ships of carbonate; calcareous algae make hardened bright pink "shrubs" in the intertidal zone; all kinds of snails and clams build shells, as do the unrelated barnacles; and some worms make calcareous tubes. After these animals die, the cast-off remains of the shells can accumulate to form significant parts of beach deposits, and those from protozoans can, over geologic time, cement together to form massive rock layers. But in life, not one of these other types organisms comes close to matching corals' scale of building.

THE CORAL'S CASTLE

The symbiotic zooxanthellae are the powerhouses of corals' construction capabilities, but the polyps do all they can to facilitate the endosymbiont's work. Scientists recently discovered that the coral skeleton is like a hall of mirrors, designed to reflect sunlight to create the most optimal light environment for the zooxanthellae. The light reflected off the skeleton is thought to at least double the zooxanthellae's energy production. That such a fundamental property of the coral skeleton could remain unknown until recently is a reminder that biological discoveries are always waiting to be made. Important truths of biology are hiding everywhere in plain sight.

Digging deeper into the superpower of building castles takes us on an interesting evolutionary journey to explore photosynthetic capabilities in corals. Not all coral species can photosynthesize, and some are more efficient than others. Acroporids are highly branched, bushy corals that stand out as being the most diverse of coral groups and may have the highest photosynthetic efficiency. While there are no doubt deeper biological issues that underlie this mastery, it is thought that the high degree of branching provides a greater surface area coupled with, on average, efficient photosymbionts and more transparent tissue. Acroporids as a group appear to have pushed the envelope toward obligate photosynthesis, with only rare cases of heterotrophy (eating small animals to supplement nutrition from photosynthesis).

This mastery will come at an evolutionary cost in the age of climate change, because the branched acroporids are usually the most susceptible to bleaching.

When scientists make discoveries about the way nature works, they often do more than increase our knowledge of the organisms with which we share the planet. Many discoveries have applications for humans. Sometimes, too, knowledge that has existed for a long time can be put to new use. In this field of bioinspired design, corals play a prominent role on at least two fronts.

Seeking to design more efficient algal biofuel reactors, researchers at Scripps Institution of Oceanography and the Jacobs School of Engineering have turned to corals for answers. They designed 3D printed coral biomaterials as templates for algal growth and found that both the normal coral algae and the bioreactor algae grew one hundred times faster on the structure of the synthetic coral skeleton than in liquid culture. Algal bioreactors now use these reflective templates instead of just vats of liquid culture.

> Full fathom five thy father lies.
> Of his bones are coral made.

Shakespeare's long-ago description in *The Tempest* of a drowned father from a shipwreck reverting to coral in a deepwater grave is becoming more real. Researchers have discovered that coral substitutes beautifully for bone in reconstructive surgery. Doctors cannot often harvest enough human bone from a patient needing reconstructed jaws, spines, or other bones. Coral bone is a more successful template for new growth than either bones from cadavers or synthetic materials. Researchers found that the organic matrix in a coral skeleton is similar in density and porosity to human bone and can support subsequent bone growth. The coral skeleton is readily colonized from nearby healthy bone by the human cells called osteoblasts, which make bone, creating a permanent seal between the human bone and coral transplant. These coral-based materials are an ideal substrate for colonization of human osteoblasts and also are inert and do not elicit an immune rejection. Furthermore, coral skeletons, which are made of calcium carbonate, can be transformed into hydroxyapatite, the major constituent in human bones, by a reaction that exchanges carbonate for phosphate.

In a further inspiration derived from the study of coral skeletons as bone implants, researchers noticed that deep-sea bamboo corals have skeletons of interdigitating bands of calcium and a softer, flexible protein. These composite skeletons can provide an extra boost to bone growth and inspired the idea of making composite-material bone implants. The continuing use of various kinds of coral skeletons to aid human bone transplants is a prominent and expanding example of the enormous biomedical value of bioinspired design.

Corals are part of the animal group called cnidarians; other members of this ancient group of predators with radial body symmetry include anemones and jellyfish. All cnidarians have harpoon-like, venom-filled stinging cells called nematocysts; soft bodies; and free-swimming larvae. As adults, some (like coral and anemones) exist as sessile polyps, while others (like most jellies) remain free-swimming (adults with the free-swimming form are called medusae). The common ancestor of all living cnidarians was a solitary anemone that lived in ancient seas as early as 800 million years ago. Colonial growth forms with multiple polyps and the ability to precipitate calcium carbonate skeletons evolved much later, between 500 million and 580 million years ago. In 2022, a newly discovered coral relative, two bifurcating polyps encased in a skeleton, was dated to about 560 million years ago and named *Auroralumina attenboroughii* for naturalist David Attenborough. One lineage of corals eventually formed their game-changing associations with photosymbionts by the Devonian period (about 415 million years ago). This symbiosis gave these ancient corals solar power and the capability to evolve rapidly and explode in number. The large and widespread coral reefs that formed during the Devonian supported diversification across all branches of the tree of life and are part of the reason the Devonian period is known as the "age of fish."

The reef-building corals are one of two cnidarian groups called coral. They have sixfold symmetry with tentacles in multiples of six, and for this reason are called hexacorals. The other coral group, which includes the sea pens, gorgonians, and soft corals, have eightfold symmetry and eight tentacles. They are called octocorals.

Although having eight tentacles sounds little different from hav-

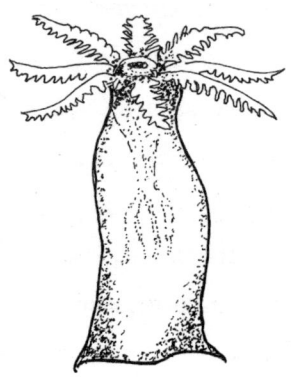

The eight-fold pinnate tentacles of octocorals.

REDRAWN FROM *ENCYCLOPEDIA BRITANNICA*, 11TH EDITION, VOL. 3, 1911

ing six, the number of tentacles reflects even more fundamental differences such as how skeletal elements are built and the shape of individual tentacles. While some octocorals form symbioses with photosynthetic algae, not all octocorals are photosynthetic, and they are not completely dependent on their symbionts for nutrition. The divide between these two groups is ancient and buried in deep time; molecular time clock studies and some fossil evidence shows they split from a common ancestor more than 700 million years ago.

On our planet, there are roughly eight thousand species of reef-building corals. Their fantastically diverse shapes range from tiny, finely branched shrubs to spreading table corals to massive towers. Corals live in well-lit and low-nutrient tropical waters and reach their highest diversity in the Indo-Pacific. The heart of coral biodiversity—a vast region in the middle of the western Pacific Ocean just north of Australia that includes the waters of Indonesia, the Philippines, Papua New Guinea, Timor-Leste, and the Solomon Islands and is appropriately called the Coral Triangle—houses upward of six hundred coral species. In comparison, there are roughly sixty-seven species of coral in the Caribbean. I was fortunate in my career to work on both Caribbean and Pacific corals. My time on coral reefs started as a graduate student studying snails that eat corals (see chapter 4) and bloomed into focused research on the inner workings of coral immune systems. On Caribbean surveys in Saint Croix, our team typically saw as many as five species of corals in one square meter. In Indonesia, there might be thirty species in a quadrat of the same size. Reefs in the Coral Triangle create a spectacular habitat for an impressively large and diverse array of fish, invertebrates, and microorganisms. There are possibly two thousand species of fish identified from Indonesian coral reefs and more than one thousand from Caribbean reefs.

Marine biologists, climate scientists, and others sounded the alarm about the health and future of coral reefs decades ago, in the 1980s. While overfishing and coastal pollution were having noticeable impacts in some areas even then, the long-term threat was climate warming. In the 1980s, climate warming was mostly an abstract, theorized problem for our future on land. But for coral reefs the implications were dire. It was known that zooxanthellae are very sensitive to temperature and would evacuate their homes inside coral polyps if the water warmed only a few degrees, causing the coral to "bleach." The zooxanthellae lend corals their color, so when they bail out, corals turn white as their skeleton shows through transparent skin. Although coral can recover if the bleaching is short-lived, bleaching can also be a prelude to death because corals deprived of their photosynthesizing partners slowly starve. In addition, scientists knew that the carbon dioxide building up in the atmosphere was being partially absorbed by the oceans and that more carbon dioxide dissolved in ocean water would make it more acidic. The more acidic the oceans, the greater the corals' difficulty in precipitating calcium carbonate and making reefs.

During the 1980s and 1990s, small increases in average ocean temperatures and acidity were recorded. These changes were difficult to detect against the background of regional and annual variation, but the trends validated climate warming predictions. In the early summer of 1997, unusually warm surface water was detected in the eastern Pacific Ocean. These warm anomalies intensified and spread over a larger area of the ocean. By September, one of the most powerful El Niño events ever recorded was in full swing. The effects extended across the globe, causing powerful storms, droughts, heat waves, and higher ocean temperatures in many areas of the tropics. By the time El Niño dissipated in June of 1998, an estimated 16 percent of the world's coral reefs had bleached and then died.

Those of us who studied corals watched these events with trepidation and warned of losses to come. We knew that El Niño was a cyclical phenomenon that occurred well before humans began to overload the atmosphere with carbon dioxide, but this time the cycle and impacts were extreme, and it seemed likely that combustion of fossil fuels had played a causal role. We received reports from around Indonesia that corals were bleaching and that small outbreaks of disease were appearing on the reefs. While the bleaching was bad news, it was expected. What worried us most was the disease outbreaks.

The multiple stresses that accompanied climate change could be weakening corals, leaving them more susceptible to attack by pathogens. Further, the growth rates of some pathogenic bacteria, fungi, and viruses increase at warmer water temperatures, creating a double danger with warming. We worried that in warming and acidifying oceans, additional killers of coral could be infectious diseases. It was time, we realized, to develop plans to monitor the health of coral reefs around the world. Only with good data on the status of reefs and how they were changing could we hope to sustain these important ocean habitats. Because the economies of many countries depended on coral reefs remaining healthy and productive, our idea was developed into a five-year coral reef sustainability program led by scientists from the World Bank and included seventy other scientists from around the world.

One outgrowth of the resulting project was a monitoring and research program we developed in collaboration with Professor Jamaluddin Jompa at Hasanuddin University in Makassar, Indonesia. By 2013 we developed a plan to survey the health of reefs from Bali to Sulawesi to Raja Ampat in Indonesia. For me it was an exciting opportunity to dive the most beautiful reefs in different parts of Indonesia with top coral biologists from Australia and Indonesia while at the same time gathering valuable baseline information on the health of those reefs.

When we survey the health of a coral reef, we can't ask a coral how it is feeling, so instead we assess visual and genetic signs of health. We score a coral as healthy if it has normal color and polyp behavior. We score a coral as sick if it is discolored or has limp, slow-moving tentacles or white spots or has lost healthy tissue and the underlying white skeleton is exposed. We also collect samples from the sick-looking corals so that back in the lab we can run follow-up genetic tests to determine whether the genetic materials from a particular pathogenic bacterium, fungus, or virus are present.

We started in Bali, near the remote island of Nusa Lembongan. At most of our sites there, the ocean races on tide changes like a fast-flowing river. Some of the currents have the dangerous property of sweeping downward like an undersea waterfall. When one of my colleagues learned I would be bringing a team, she cautioned me that several people had died in those currents that year. The high currents create dangerous diving conditions but also support sparkling reefs filled with hundreds of fish species, banded sea snakes, bright speckled nudibranchs, striped sea feathers, and a multicolored coral foundation supporting it all.

There were six of us, three teams of two divers each, on a dive on Bali's Ceningan Wall. One team would identify and count fish, and two teams would count all the corals on three transect lines, each thirty feet long. We started the dive hanging in the current on ropes attached to our boat until everyone was ready. As a team, we all descended fast and fought the current to arrive on the bottom in the same place. As I hung underwater, holding on to a rope underneath the boat to stay in place, I felt like a sheepdog watching my flock. I was not only dealing with my own challenge of staying oriented in the current but also watching as my teammates reached the bottom, laid their transect lines, and got to work. So it wasn't until I finally settled to the bottom on my transect that I could look around and feel the magic of being in the midst of the most diverse coral reef in the world.

I was swept through a menagerie of color and dazzled by an unusual blue hard coral that was often nestled against a bright orange soft coral. Above the phenomenal kaleidoscope of coral on the bottom swam many kinds of fish, big sea snakes, and tiny bright blue moray eels, with five different species of clown fish stationed in their multicolored anemones. In the current, the swaying coral and anemones ranged from orange to blue to pink to brown and maroon, the hues all merging in Van Gogh–esque whorls of color. On these reefs, the variation in the branched forms of coral is staggering.

The easiest way to imagine it is to think of the variation in the branching patterns of trees and shrubs and the range of flower colors you see in a formal garden and then to picture each tree and shrub spun in a fine, delicate lattice of multihued concrete.

I saw about twenty different species of branching corals just within the first fifteen feet of my transect. The strange bright blue corals surprised me. Blue is the rarest of natural colors because it has a very short wavelength and there are no natural blue pigments. Then I noticed something odd about the structure. In the end, I realized that it was an impostor: on the surface, it was hard and looked like a reef-building coral, but it was actually a hydrocoral, in the same group as the fire corals. It is not a trivial difference; the hydrocorals are coral relatives that do not have the superpower to photosynthesize like the reef-building corals. Instead, they have powerful stinging cells. When you touch a fire coral, hundreds of nematocysts penetrate your skin and inject a toxic venom. It feels like a bee sting at first, but the pain increases over time as the toxin reacts. The welts that develop take a week to heal and another week to stop itching. Once we knew what they were, we gave the pretty blue corals a wide berth, even though our full wet suits provided good protection.

There were also at least three species of the large branching brown acroporid corals, towering above the reef in five-foot-high branching spires. There were bizarrely branched pectinid corals, shaped like ice cream cones and with teal-green edges and bright stripes radiating on a brown body. There were at least ten species of short, very delicate, finely branched corals growing as an understory in the interstices of the larger corals. There were rounded brain corals. Looking a little deeper than our transects, I could see plating corals stretching fifteen feet across.

Even in the heart of coral biodiversity, corals face the existential threat of climate warming. Some of the corals we surveyed on those reefs in 2013 were in poor health. Warming events had broken apart the delicate symbiosis with the zooxanthellae and activated infections that stripped the live tissue from the skeleton and killed the entire colony. We were seeing more evidence that the declines of coral reefs that started with the worldwide coral bleaching event in 1997 and 1998 had continued. Indeed, all the coral reefs we have studied, from Saint Croix to Mexico to Australia to Palau to Zanzibar, have at some point in the last decade turned white in bleaching events caused by increased temperatures. The zooxanthellae are at once the heart of the coral's superpower and its Achilles' heel.

Unfortunately, one of our big discoveries from the Indonesia survey was that it wasn't just heat pushing coral susceptibility to infectious disease. Local factors like plastic, dynamite fishing, and sewage pollution were breaking and stressing corals, so infections could easily start. As we swam our transect lines and checked each coral at sites from Bali to Raja Ampat, we found a lot of plastic entwined in the branches of living coral. On reef after reef, we noticed plastic trash. Old fishing line, chip bags, diapers, and single-use plastic bags were surging into the ocean from coastal towns and rivers like some school of alien, nonliving fish, and they were ending up tangled on corals.

My postdoctoral scholar, Dr. Joleah Lamb, now a professor at the University of California, Irvine, had discovered that plastic fishing gear caused coral disease in fished areas but was reduced inside of marine protected areas. She noticed that the plastic was creating abrasions on the coral skin and that these abrasions were often the sites of infection. She proposed that we tally the plastic in our health survey. I was opposed at first, because it was already a hard survey to complete, but she talked me into it. When she combined our data from Indonesia with surveys she had done in Thailand, Myanmar, and Australia, the results were staggering. Plastic entangled on corals was increasing the risk of disease by 85 percent.

As part of our wider coral survey, we taught a workshop on coral health in Indonesia, two hours across the Makassar Strait on the tiny island of Barrang Lompo in the Spermonde Archipelago, Sulawesi. The goal was to teach the basics of field surveys for monitoring the health of the corals. It was the beginning of a ten-year collaboration with Indonesian scientists, primarily Dr. Jamaluddin Jompa and his students and colleagues at Hasanuddin University. It was special to run a workshop like this in Indonesia, because this is the heart of the coral triangle, with the richest coral biodiversity on our planet. Indonesian scientists from across the country attended, because they were only too well aware of the extreme economic value of the services provided by coral reefs and the rich biodiversity housed on them. Top coral experts in the world joined our team to teach in the workshop. Professor Bette Willis, from James Cook University in Australia, and one of Professor Jamal's former professors, taught a series of lectures on how to identify different coral species from their skeletal form while diving on underwater transects. Professor Laurie Raymundo trained as a PhD student with me at Cornell and was then a leading researcher at the

University of Guam, developing key methods to restore coral reefs. Courtney Couch was a coral health PhD student in my lab (and also joins us in chapter 5), and Dr. Erin Mueller was a postdoctoral scholar working with us. Laurie and Bette also gave lectures on newly described health syndromes they were discovering on Pacific corals. We ran this workshop at the remote marine lab in the Spermonde Archipelago so that we could teach lectures in the morning and practice fieldwork skills like coral species identification and diagnosing coral health conditions on dives in the afternoon. We ran a whole series of sessions about detecting coral disease and its impact on the health of the whole ecosystem. While we didn't teach about human health, we knew waterborne pathogens could cause human diseases in coastal ecosystems. Little did we initially realize how intertwined are human health and environmental health. We were about to experience a real-life lesson.

The three-day workshop went well, and we were pleased with what we had accomplished during both the indoor lecture sessions and field-based reef surveys. On the last day, two of the Indonesian graduate students never came to breakfast, and when Courtney went to check on them, they simply said they were tired and not hungry. I plunged in for a hearty meal and savored the strong Indonesian coffee. Laurie, Bette, Courtney, and Erin were all in high spirits. We had a sense of hope that our science would contribute to improved management of the beautiful reefs we had fallen in love with, and we looked forward to more collaboration with Indonesian scientists. We packed up and boarded the boat. About halfway across on the hour-long ride to Makassar, our spirits dropped. By the time we reached

Makassar, I was tired, light-headed, and nauseated, and I barely made it to the public bathroom at the port.

Things quickly went from bad to worse. I had to be helped into our hotel. I lay in my room going in and out of thought, honestly wondering if I would make it through the night. In the morning I was able to sit up, but my teammates in the next room were now all sick. Over the next two days we managed to get everyone to a clinic in Bali. We all had amoebic dysentery, caused by *Entamoeba histolytica*. This rather horrifying pathogen literally eats human gut cells alive, piece by piece. A two-week course of the unpleasant medication Flagyl brought us all back, except for Dr. Bette Willis. She grew weaker by the day. It turned out that, in addition to dysentery, she had typhoid. She had to be flown immediately home to Australia, where it took her months to recover. While we all recovered from that episode, I still have lingering effects from the various amoeba infections I picked up over the years in Indonesia.

Having my whole team come down with dysentery on Barrang Lompo was a strange aha moment for me. A light went off in my head about the visceral reality to the idea popularized as One Health, that human health and environmental health are inextricably intertwined. We had come to study how coral health was compromised by diseases, only to learn that in a sewage-polluted environment our own health could be endangered. It was a call to action for me to explore the links between poor coral health and poor human health. This would lead to over a decade of projects connecting human health and ecosystem health in waters from Indonesia to Hawaii and finally Puget Sound, in Washington.

Our work on the most beautiful coral reefs of the world continued to include disheartening experiences with severe plastic pollution and blast fishing with dynamite. Additionally, we were staggered by the horrible impact of sewage pollution, a seemingly manageable threat to ocean health. We were weary of seeing oceans as victims, and we were traumatized by the potential loss of such an immense portfolio of riches from wave protections onshore to new bone therapies for humans to frontiers not yet discovered. We wanted our research to show how ocean biota creates solutions. We developed a new initiative in the second phase of the World Bank project to explore the valuable ecosystem services that might be lurking in the splendor of coral reefs and in associated habitats like seagrass meadows and mangroves.

Joleah and I talked over possible benefits of tropical habitats and pondered our earlier realization that the health problems of humans and corals are linked by the same poor water quality. We decided to find out what was in the water at Barrang Lompo and then identify a property of the ecosystem that could help clean it up. We started with a simple measurement of the level of human fecal bacteria in the water. It was terrifyingly high and a clue to why we all became so ill: more than 2,000 cells of culturable bacteria per milliliter. The EPA's health limit for beach closure in the US is 125 cells per milliliter. But then we took enough samples to see a pattern emerge: levels were drastically lower in the seagrass meadows surrounding the island. In collaboration with Professor Jompa and his students, we cultured water around four islands and detected the same pattern at all sites: high levels of fecal coliforms near the beach that slowly dropped off by dilution into deeper water. But the levels dropped twice as fast in the seagrass. We knew then we had the right project for this World Bank program: to measure the ability of seagrass meadows to clear bacteria from water. Joleah went on to take more samples and completely sequence all bacteria species in the water surrounding the four islands. The results were clear: a huge drop in levels of known pathogenic bacteria in the seagrass at all islands. To complete our One Health vision and show that human health and coral health were closely linked, she showed that coral disease was also drastically reduced in the seagrasses.

We hope this work will propel needed changes in both sewage treatment and conservation of valuable seagrass meadows. While we don't claim that seagrasses alone can handle huge loads of human sewage, we do contend that coastal vegetation can reduce the constant spillover sewage that is not handled well by existing sewage systems. This points to a clear policy objective: take steps to preserve coastal seagrass meadows and, where possible, increase their extent.

We went on to record high levels of fecal coliforms (and demonstrate the need for better septic management) on the Big Island of Hawaii. Next, we set our sights on Puget Sound and the city of Seattle. Even though Seattle is huge, favorable water currents support lush seagrass meadows along its shores. Could seagrass make a difference in dangerous bacteria levels in these US urbanized waters? With the help of Seattle's Mussel Watch program to outplant mussels as bacterial concentrators, we were able to show that some dangerous pathogenic bacteria were lower in mussels from seagrass meadows than in adjacent waters. Seagrass meadows and other kinds of coastal forests and mangroves, are a powerful natural detoxifier of pathogenic bacteria wherever they occur. Some have called these areas of coastal vegetation the kidneys of the planet. I continue to research the ways seagrasses detoxify bacteria, and push hard for conserving these unusually valuable marine ecosystems.

Despite our hope to focus on positive solutions, some days it felt like a game of Whac-A-Mole. As soon as one problem is solved, another rises up. While we were documenting the hygiene services of seagrass ecosystems, climate warming was continuing to threaten coral reefs and even chipping away at the health of seagrasses. Habitats beneath the waves were changing more rapidly than we could measure.

After the coral bleaching event in 1997 and 1998, coral reefs worldwide continued steep declines. In 2008, following multiple catastrophic bleaching events, the International Union for Conservation of Nature declared one third of all coral species at risk of extinction. While any extinction is a tragedy, the loss of coral species at this scale is catastrophic because so many other species are dependent on coral reefs. It is extremely difficult to confirm extinction in the ocean, but we have undoubtedly already lost species of reef-building coral from our planet.

The best-documented impacts of warming are on the Australian Great Barrier Reef. A study in 2021 showed that coral bleaching has affected 98 percent of the Great Barrier Reef since 1998, leaving only a tiny fraction of reef unimpacted. Dr. Terry Hughes, former director of the Australian Research Council Centre of Excellence for Coral Reef Studies, based in Townsville, reports that the five bouts of mass bleaching since 1998 have turned the Great Barrier Reef into a checkerboard of zones, with different impacts ranging from minor to severe. Eighty percent of the reef has bleached severely at least once. Dr. Hughes reports that the long-term outlook for this 1,400-mile-long ecosystem is poor, even though it is arguably one of the world's best-managed coral reefs. Ironically, while a good steward of the world's most stunning underwater seascape, Australia is also one of the world's biggest exporters of coal and gas, with an economy heavily reliant on fossil fuels. Australians are therefore contributing to the annihilation of their own reef.

Of the dangers posed to ocean biota by climate change, higher water temperatures have thus far caused the most mortality on coral reefs. But increasing ocean acidification looms as a fatal threat in the long run because it impairs coral's ability to precipitate calcium carbonate and build robust skeletons. To add to the problem, the rapidly rising sea levels that accompany warming can outpace a coral reef's upward growth and reduce its access to required sunlight.

Scientists look to coral's powers of regeneration and evolutionary adaptation for solutions. The hunt is on to find species and strains with unusual capacity to adapt to warming and the fresh onslaught of heat-fueled microorganisms. Some populations of coral may already possess genetic variations that make them less vulnerable to bleaching or better able to recover after a bleaching event. Research is underway to identify coral and algae genes that are more resistant to bleaching and to potentially engineer more resistant coral strains. As I discuss in the epilogue, more-resistant corals could be seeded into protected areas where other corals have died.

It's a hard fight, however. The current rate of warming is unlike any experienced in the past 66 million years, as noted in the *Sixth*

THE CORAL'S CASTLE

Assessment Report of the International Panel on Climate Change. Humans and corals are now both in an existential race to adapt to climate change. As a marine scientist, I am keenly aware that many solutions and mitigation strategies involve our oceans. The oceans already help us out by absorbing 25 to 30 percent of our greenhouse gas emissions. Scientists and engineers are looking to several additional ocean-based strategies: harnessing the energy of waves and tides, deriving so-called blue carbon from oceanic vegetation, storing carbon in the seabed, and shifting to ocean-based food instead of intensive land-based agriculture. As we humans barrel down this path toward rising seas, extreme weather, and heat that tests our planet's limits, some of us will also be watching underwater to see how the oldest creatures adapt and what new powers emerge.

Gorgonian soft corals come in a diversity of colony forms, from encrusting to thin and thick sea whips to sea plumes and sea fans.

3.

The Sea Fan's Ancient Defenses

Each day during breakfast, from our bunkroom window sixty feet below the sea's surface, we watched the daily cycle of life begin on the reef. The scene was set by towers of pillar corals rising above the cliffs, hilltops, and valleys created by a rainbow of red, orange, brown, green, and yellow corals, interspersed with bright sponges. As the early morning light filtered onto the reef at sunrise, green-blue parrotfish, large rainbow parrotfish, yellow and black tangs, and a stream of smaller, multicolored fish all left their nighttime perches in crevices and caves and began their early morning commute into food-rich waters.

About twelve hours later, at day's end, the light dimmed slowly underwater, and all kinds of fish hurried back to find their spots on the reef before dark. It was the day parade in reverse as parrotfish, schools of tangs, and the big green rainbow parrotfish retraced their fin beats home. The now-endangered rainbow parrotfish

(*Scarus guacamaia*), some of them up to two and three feet long, would sail by in small schools. When not out diving, we would sit by the window and watch the show, precisely synchronized to the sun and moon as if each fish consulted its own internal timepiece. We felt the ancient drumbeat. Light and tides have driven this daily cycle of life on the reef for millions of years.

I was starting a week at sixty feet under the surface in a very confined space with three other scientists, and it ended up being the adventure of a lifetime. It was 1983 and I was still a graduate student, the junior member of the team, feeling very lucky to have been invited along as one of the expedition's divers. Our leader was Dr. Tom Suchanek, then a resident researcher at West Indies Lab in Saint Croix. Dr. Bob Carpenter was an up-and-coming coral reef biologist, also working at West Indies Lab. Dr. Jon Witman, from the University of New Hampshire, was an experienced underwater photographer and cold-water diver and marine ecologist. We were together immersed in the wonder and adventure of living underwater, exploring and making new discoveries on a rich Caribbean coral reef.

After we finished our lasagna one evening, the four of us donned our wet suits, masks, and fins and radioed to the surface support team that we were leaving Hydrolab, our underwater laboratory and bunkhouse. We dropped through the open hatch inside the habitat and swam underwater to a radio waystation, sitting beside the habitat and equipped with a radio and air supply. After we radioed an update of the evening work plan to the surface, we swam the twenty feet to four sets of twin-80 scuba tanks strapped to the railing. The air was already turned on so we could grab the regulators and breathe as soon as we reached the tanks. Although they were heavy to lug around on land, the twin 80s were buoyant underwater and felt lightweight to maneuver. As we headed out, I looked back at the spectacle of the lighted underwater habitat, sitting like a spaceship on the bottom of the ocean at fifty feet, attached by a huge tether to a surface boat that pumped down breathable air. Hydrolab was like a giant pipe with four legs perched on the bottom. Attached was a tank rack and waystation containing compressed air and a radio to the surface. Light shone out of the main picture windows in the dining and sleeping quarters and through our entry port, like a safe cabin in the woods in comparison to the dark reef ahead.

Tanked up, we started the long swim to our worksite on the reef drop-off. Although the swim was only fifteen minutes, it seemed much longer—time stretches underwater in the dark with heavy dive gear. We were checking on experiments we had set up two weeks before. We swam tightly as a group of four, safety in numbers our only defense against anything unexpected on the night reef. It felt like the dark reef watched us, silent and still in our flashlight beams. We didn't see eyes, but the wakeful residents of the reef could see us. The darting fish that had been out during the day were sheltering in lobes and crevices under coral heads. We stopped to look at a parrotfish, snoozing under a coral head and encased in its sheer mucous sheet like a sleeping bag. It's surreal to see a normally fast fish sleep—still, and with eyes open. They look awake lying on their rock or coral shelves, but are sluggish and unresponsive, only slowly awakening if you pick them up. Since one can't talk underwater, we communicated our wonder with bubbly *hmm*s, shaking our heads in amazement and meeting expressive eyes through masks. As we searched with underwater lights, we saw more blue-green parrotfish and red squirrelfish tucked away on ledges for the night.

We continued along the reef, flashlight beams illuminating a previously established rope path. This was one of my first night dives, and I was nervously swinging my light in an arc ahead of us, for fear bigger things might come our way. This reef was perched on the edge of a huge drop-off into deep water and the open ocean, so oceanic fish like sharks could be cruising nearby in the dark. I didn't like the idea that a shark could easily trail us, unseen in the night. Most of the dive was quiet, still, dark, and eerie. Suddenly, I saw a flash of movement coming fast toward us, and my fears about deep-sea monsters surfaced. I yelped and Tom laughed, flooding his mask. It was only a midsize green turtle, the size of a dinner plate, zinging at high speed out of the dark and careening past our heads, probably attracted by our lights.

It's ridiculously easy to get disoriented on a reef at night. It's hard to even know which way is up because the pull of gravity does not work underwater and the darkness is total unless moonlight filters down from the surface. We needed markers to find our worksites at night and our way back to our habitat. At our research site, on the edge of the deep slopes of Salt River Canyon, Saint Croix, there was a network of rope pathways anchored to the ocean bottom with attached metal arrows, all pointing the way to the habitat. Imagine your mask being knocked off or flashlight lost in a big current and you would need to find your way back to the habitat and safety by holding the rope and feeling the arrows.

If one did manage to get lost, surfacing was not an option. We

THE SEA FAN'S ANCIENT DEFENSES

were saturation diving, staying underwater for five days. The concentrations of the gases in the compressed air we breathed from our scuba tanks had long since equalized with the levels in our body tissues, which meant that our bodies were full of tiny nitrogen bubbles. We were like champagne bottles: if we left the high-pressure environment and surfaced suddenly, we would blow our corks. The bubbles of gas would expand and fizz out of our tissues, lodging in our joints, spinal cords, and brains. The painful, sometimes lethal symptoms of these expanding bubbles caused by rapid surfacing is called the bends, or decompression sickness. So, no matter what happened on the reef, we had to stay at depth, even though it would take only two minutes to swim up sixty feet to the surface. The underwater habitat was equipped to allow us to stay safely on the bottom for an entire week. At the end of our stay, in the locked-down habitat, we would spend thirty-six hours decompressing, slowly returning to surface pressure as the gas fizzed harmlessly out of our bodies. This evening, after following a rope for fifteen minutes to the edge of the deep reef, we finally reached the spot where we would watch the brutal territorial battles among corals unfold.

Corals live like Dr. Jekyll and Mr. Hyde. During the day they reside peacefully on the reef, allowing their symbiotic algae to absorb sunlight and make food. At night, the Mr. Hyde side emerges. They become combative and confrontational; vicious, slow-motion battles erupt between neighbors. These fights are the outward signs of an ongoing struggle for space. Healthy coral colonies are always expanding as the individual polyps clone themselves, or "bud off," to form new polyps. This expansion soon creates conflicts with neighboring corals, as well as with other sessile organisms, like sponges, as all of the polyps must be attached, directly or indirectly, to the reef. The corals are well equipped for the fight. After dark, polyps extend extra-long tentacles studded with venom-filled nematocysts. The harpoon is released with great force when the nematocyst comes into contact with living tissue. So any neighbor within reach of these tentacles gets shot full of poison harpoons. A coral fight can be vicious and to the death, but it's never quick. Nematocyst stings stun a polyp and damage tissue, but it takes many stings to kill one. The battles can be slow to start and long to finish; some continue for months.

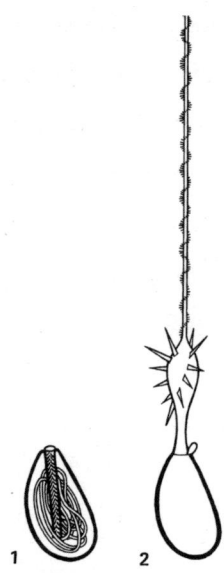

Undischarged and discharged nematocysts, separated from a surrounding nematocyte cell. (1) An undischarged nematocyst showing the coiled tube. (2) The discharged nematocyst showing barbs protruding from the tube.

DIAGRAM COURTESY OF VICKI PEARSE, *ANIMALS WITHOUT BACKBONES* (CHICAGO, IL: THE UNIVERSITY OF CHICAGO PRESS, 1987)

THE SEA FAN'S ANCIENT DEFENSES

A few weeks before this night dive, we had uprooted several purple gorgonian soft coral colonies (*Erythropodium caribaeorum*) and transplanted them near some medium-sized green, stony lettuce corals (*Agaricia agaricites*). We wanted to see if the resident lettuce corals would defend against the new neighbors' intrusion into their space. Transplanting a coral is sort of like propagating a new plant in a garden. You break off a fragment and cement it with underwater epoxy onto the living reef; it will "take root" and grow by adding new polyps at the edge.

This evening, in the glow of our flashlights, we found our metal markers and the green lettuce corals with new purple neighbors alongside and looked closely at the borders between them. The lettuce corals, with all their tiny polyps expanded at night to snag any wayward zooplankton, sparkled in the lights. Normally their polyps are very small, with tentacles only a few millimeters long, but now, in the region of the coral adjacent to the new invaders, extra-long tentacles with bulbous ends stretched two inches across the gap. Our experiment had triggered the expected response: the resident corals armed for battle with extra-long, harpoon-laden tentacles called sweepers only three weeks after detecting an intruder. It was life-changing to see this firsthand—an elemental and yet very specific battle for space by some of the world's most ancient creatures that had rarely been observed by humans. This was a measured, aggressive response induced by proximity to a competitor for space. We now knew exactly where the coral activated sweepers for battle, against which intruder, and how long it took.

The sweepers rippled in the current, extending across the gap between the coral colonies to brush against the purple gorgonian, triggering the tiny harpoons to fire and punch micro-wounds into the enemy coral's skin. When the attacked newcomer took a hit, its own tentacles retracted. A patch of skin on our transplant was wounded, torn, pale, and discolored. The gorgonian was injured and slow to respond, but it was still in the game. In two more weeks, we would return to see that the gorgonian had grown its own sweeper tentacles in retaliation, each packed with venom-filled tiny harpoons. Over weeks to months of battles like this one, the final winner in the fight for space would be determined by factors such as the size of each colony, its ability to use solar power, and how much energy each had to put into the fight.

Our experiment with the purple gorgonian and the green lettuce coral was part of a larger project that Tom had developed to learn how reef-building, or hard, corals were holding their territory on a crowded coral reef. He wanted to see if the hard corals might be losing space to competitors like sponges and soft corals. He had noticed that corals were fighting not only with one another for space on the reef but also with sponges and other encrusting animals that could send out advancing tendrils, sometimes laden with toxic chemicals. In this way, other encrusting animals could kill corals and take over their reef space. Because corals are the principal reef builders, Tom wondered what factors might be giving sponges, sea squirts, and soft corals an edge in the fight for space against hard corals. As beautiful as these other animals are, they do not build reefs that can sustain and nurture the lives of so many fish and other animals. Was diminishing water clarity, caused by increased nutrient pollution from humans on land, handicapping the solar-powered corals? Did the attenuation of light at greater depths disadvantage the corals living there? The location of the saturation diving laboratory on the edge of a deep reef gave us access to extreme research areas and the chance to investigate how water depth, as a proxy for light penetration, might affect a coral's ability to fight for space. The presumption was that corals would be most successful in fighting near the surface, where their solar power was strongest.

Tom, Bob, John, and I worked on reefs at 10, 60, 90, and 120 feet below the surface, day and night during the week of our saturation mission, using photographs to map the encounters between corals and sponges, taking photos all along the marked area, and returning a year later to photograph the exact same locations. Comparing the before-and-after time series showed that sponges were aggressive competitors for space with corals, particularly at depths of 90 and 120 feet.

The week I spent working on the reef around the Hydrolab in 1983 was deeply formative. As an ecologist, I was fascinated with the competition for space on reefs. But as I observed the outward manifestations of that struggle—like the fighting tentacles that developed on the threatened lettuce coral—I couldn't help but wonder what was going on that I couldn't see. How did the resident lettuce coral know that the polyps of the transplanted gorgonian weren't part of its own colony? Earlier experiments had shown that a coral will not fight with itself, even if a piece is removed, cultivated in the lab, and transplanted to its border. Sweeper tentacle growth is triggered only when other coral species or nonself-genetic types of a coral's own species are detected nearby. At its heart, this was a recognition process. How did it work?

Pondering this question helped me realize that the boundary scientists had set up between ecological processes like competition and the biochemical processes of immune systems was an artificial one. Defending against the threat of being pushed out of your space and defending against the threat of being eaten from the inside by microorganisms are, I realized, different aspects of the same kind of problem. And, as I discussed in chapter 1, responding to both hinges on being able to recognize and respond to nonself. So watching the battle for space on the reef and the recognition process that was a fundamental part of it got me thinking a lot about immune systems. It was mind-bending to realize that corals and sponges have one of the most ancient immune systems, and here they were, putting a 600-million-year-old process to work in modern waters.

Another lasting consequence of the week of saturation diving research around the Hydrolab was my deepened appreciation of gorgonian corals. Gorgonians are relatives of reef-building corals, members of the octocoral group that I described in chapter 2. Unlike the reef-building hexacorals, they do not create hard skeletons and do not always have photosynthetic algae living inside them. Gorgonian corals are distinguished by their polyps' eightfold symmetry and stiff but flexible internal protein skeletons. The different forms in which they occur have inspired the common names sea fan, sea whip, and sea plume. They live on reefs around the world, but are most diverse on Caribbean reefs, with thirty-nine species known from the region. In some Caribbean locations, the gorgonians are even more diverse and abundant than reef-building corals.

THE SEA FAN'S ANCIENT DEFENSES

As you will see in chapter 4, as part of an independent research project that I carried out while living aboard the Hydrolab, I spent much time that week looking for gorgonians that had been attacked and partially eaten by a predatory snail.

Three years after our saturation mission in Salt River Canyon, I was hired in 1986 as a new professor in invertebrate biology and marine ecology at Cornell University. A larger world of opportunity then opened for my underwater research. The puzzle of how corals defended themselves against biological threats—other corals and predators and pathogens—continued to fascinate me. I also thought that understanding what was effectively one of the earliest immune systems was an important avenue of research. I was drawn to the frontier of understanding the remarkable adaptations corals brought to underwater arms races and how their ancient immune systems, still effective after 600 million years, protect against modern threats. They must have secrets to share in the fight against disease-causing microbes, I reasoned; secrets that would be useful to those of us on land. Time would show that I chose these research interests well. The changing climate of our planet was strengthening the foothold of pathogens in both the oceans and on land, shining a light on how oceanic animals and humans alike fight disease.

We had noticed exciting patterns in octocoral aggression and defense against predators during our saturation diving missions and, as described in chapter 4, chemists who searched for medically useful compounds in sponges were finding potent chemicals in octocorals. Thus, I began researching the immune systems of my old friends the gorgonians, and my lab continued this work for over two decades.

All forms of coral carry chemicals that are considered bioactive because they are unusually potent and disrupt basic biological functions, like cell division or inflammation. They can reduce inflammation in humans or stop bacterial and fungal cells from growing and even poison fish. Octocorals, along with sponges, take this ability to make bioactive chemicals to unusual levels—they are notorious among animals for the diversity of strange and potent chemicals they produce. These bioactive molecules are termed "secondary" chemicals because they have no known function in primary metabolism; they are similar in this way to plant or sponge defensive chemicals like diterpenes and phenols. But, unlike plant chemicals on land, secondary chemicals become even stranger in the mineral-rich ocean water and can be studded with toxic elements like chlorine and bromine. Little is known about the relative roles of the gorgonian itself and its symbionts in the manufacture of some of the secondary compounds. Some scientists speculate that the octocorals have taken an evolutionary path different from that of hexacorals: instead of building massive calcified skeletons, they have innovated spectacular chemical defenses to fend off all kinds of biological threats. Octocoral chemical wizardry is clearly tied up with what we think of as defense and immune function even if some of their chemical defenses are not directed against pathogens per se.

The inner workings of gorgonian immune systems were poorly known when I started as a young ecology professor at Cornell. Un-

THE SEA FAN'S ANCIENT DEFENSES

locking secrets of coral immunity required chemical instrumentation and molecular biology know-how that I lacked and, in some cases, had not yet been developed. I needed help from expert chemists, and I was fortunate that those chemists thought ecologists had useful insights into the reasons for biologically active chemicals existing in nature. Our early work in the Caribbean focused on questions that began during the saturation diving research in 1983: the fights for space among corals and the chemicals in corals that deterred fish and snails. Nature was about to push our work to a new level in confronting an existential fight between octocorals and pathogenic microbes.

In 1992, a colleague emailed me that sea fans were dying from a mysterious disease in parts of the Caribbean. Five years after our research on the chemical defenses of gorgonians in the Caribbean had begun, our work on gorgonians suddenly switched from looking at defenses against big predators like snails and fish to understanding the defenses gorgonians deployed against microbes in a deadly, Caribbean-wide outbreak of disease.

I flew in midwinter with my postdoc, Dr. Kiho Kim, from cold and snowy Ithaca to San Salvador, a remote island in the Bahamas. We launched right away in the afternoon and motored to our field site, landing on a deserted beach of white sand. Kiho and I just sat on this warm, welcoming beach for a while, surrounded by our dive gear, transect tapes, collecting bottles, and underwater data slates, talking about what we would look for on the small reef just out from the beach. What did we expect to see, and what would our sampling strategy be? Our goal for the two-week expedition was to see what the outbreak of disease in sea fans looked like and whether we could develop a research project around it. How many sea fans were affected? What did a sick sea fan look like? Were young and old sea fans equally affected? How fast was the disease progressing? Were some parts of the reef hit harder than others? These were the kinds of questions I had in mind as a practiced observer in nature, the kinds of questions that would lead us to understand the biological processes at work behind the patterns that we could measure. But we were in uncertain waters—very few marine biologists had watched in real time the spread of a new underwater epidemic. Despite the planning, we were not prepared for the underwater scene we experienced.

It started as a normal dive in warm, clear water on a still day. We swam first through the surf and shallow back lagoon, kicking along above mesmerizing ripples of brown and tan sand to reach the reef. In deeper water the first coral heads appeared. The hard corals and sponges on the reef edge looked healthy, and fish of all sizes darted and swam around us. A rather large, four-foot-long barracuda patrolled along with us, teeth gleaming when it clacked its jaw in a territorial display. Then we reached the zone of sea fans, a virtual forest of hundreds of purplish-blue sea fans taller than me, intermixed with lemon-yellow sea fans, all swaying like a field of corn in the wave surge. We started seeing both purple and yellow fans covered in vivid purple spots with gaping holes in the internal mesh. In a mental picture I carry to this day, in one hollow of the reef, a vast graveyard of over fifty sea fan skeletons stood, giving silent and eerie testimony to the biological disaster that had befallen them. That day changed the course of my research.

Over the next two weeks, we marked (with numbered metal tags) and photographed colonies so we could watch how fast the lesions on some of the sea fans expanded. Some lesions grew so fast that a colony would be dead on our return a week later. The pace of death and reach of the epidemic was unsettling. We saw in San Salvador a scenario that was playing out across the Caribbean. Millions of sea fans were dying in this epidemic. But there were survivors: sometimes the lesions on our marked colonies would stop growing, disappear, and be replaced with regrown tissue. This suggested that some gorgonian corals were successfully fighting back the unknown pathogen with a strong and effective immune response.

Watching the rampage of the infections, like leprosy on a sea fan, showed me in a visceral, holistic way the existential danger of underwater pathogens. But more important, the epidemic gave me the chance to ponder the other side of the arms race—the role of invertebrate immune systems in battling disease-causing microorganisms. I had already been worried that, in studying how chemicals produced by gorgonians deterred predators, we were on the wrong track toward understanding what was driving the evolution of chemical defense in these animals. We now saw that something in the gorgonians was stopping the growth of—or even killing—the pathogen causing the epidemic. This raised a big evolutionary question: Might microorganisms, rather than predators, have been the selective pressure behind the origin of gorgonians' (and sponges') chemical defenses hundreds of millions of years ago?

That's the kind of question one could spend a lifetime trying to answer. In the meantime, the outbreak raised questions that we could begin to answer right away. How were the gorgonians fighting the infection? Was the pathogen a bacterium, a virus, or a fungus?

Where had it come from? Why hadn't it caused widespread disease before? Leaving these questions aside or to other experts for the moment, my students and I went to work on what we saw as the most basic first step: simply documenting the scope of the outbreak. Using standard epidemiological practices, we took data on individual sea fans by marking with tags and photographing thousands of them in the Bahamas, Florida, and Mexico's Yucatán Peninsula. Then we counted. At many of our sites in Florida and the Yucatán, more than 90 percent of the sea fans died over the time of our survey. Some of these corals were over fifty years old and so had weathered storms and predation on these reefs for decades, only to all die catastrophically at once. In the Yucatán, near the town of Akumal, we had access to both shallow and deeper reefs, so we marked individual sea fans at both depths, thinking perhaps the deeper ones would survive better. On one trip to study the sites fifty feet deep, we calculated that a quarter of the sea fans had already died, and we saw and mapped so many sick sea fans that I feared all the rest would die too.

In the summer of 1996, four years into the epidemic, Dr. Garriet Smith and Dr. Kim Ritchie, microbiology professors at the University of South Carolina, identified the sea fan killer as a fungus called *Aspergillus sydowii*. It was a surprising discovery, because *A. sydowii* is typically a terrestrial fungus that is known for infecting the lungs of immunocompromised humans, often killing them. Pinpointing the causative microorganism in sea fan deaths was key to unlocking tools to study the animals' immune response and making our work relevant for human health. Could we discover novel chemicals or a different immune system response of sea fans that could be used by humans fighting fungal diseases? Or to think about it the other way, because this fungus is a disease of immunocompromised humans, were the sea fans experiencing immune compromise also?

To get us started in our own lab work, Garriet taught us basic skills in microbiology, including how to isolate the fungus from sick sea fans and to grow it under sterile conditions in petri dishes. At one point, we had a sort of fungus farm in my lab at Cornell, an entire culture chamber full of dozens of petri dishes with strains of *A. sydowii* from around the Caribbean. When we infected sea fans with the fungus, they first developed bright purple spots that grew into lesions, which grew into holes in the mesh. The holes were surrounded by brilliant purple halos. We noticed that, on some sea fans, the lesions stopped growing within the purple halos and then the wounds recovered. What was in those purple halos?

To test the strains and activate the immune response of the sea fans, we injected a calibrated slurry of the fungus into healthy sea fans in seawater tanks at the marine lab on San Salvador and later in my lab at Cornell and then watched the development of the purple halos on the surface of the fans while we tracked the cellular and genetic changes inside the tissues. Ours was the first study to investigate, in this holistic way and in real time, how the immune system of an extremely ancient animal, like a coral, responds to a particular pathogen. This was what I had been seeking for years—a way to study the raw fight between pathogen and coral immune system. Our ability to do those experiments was a lucky break because, even to this day, it is rare for ocean researchers to have a disease agent that is easily cultured plus a host that can be grown well in the lab. We had both, plus a strong team of undergraduate, graduate, and postdoctoral researchers. We took full advantage of the opportunity to run dozens of experiments in lab and field sites from the Bahamas to the Florida Keys to the Yucatán Peninsula that eventually resulted in more than thirty scientific publications.

In these experiments, the coral defenses unfolded before our eyes. The purple halos developed around five days after the infections. Under a microscope, we could see on our slides the immune cells in the skin of the sea fan responding to the fungus. Postdoc Laura Mydlarz (now a professor at the University of Texas, Arlington) created a series of histological slides that showed the emerging immune reaction inside a sea fan. Laura had come to my lab as a young scientist with training in chemical ecology and experience in the world of research focused on coral drug discovery. Her slides revealed not only that the specially stained immune cells proliferated and clustered around the infection site, but also that the immune cells were carrying and laying down tiny black granules of the protein-like chemical, melanin. These granules aggregated to form dark melanin walls. These walls appeared on the slides to act as an impenetrable barrier, trapping the tendrils of the fungus and preventing them from attacking the coral polyps. This was it: the action of the coral fighting back inside the purple halos. The purple halo itself was caused by a combination of the melanization reaction and the overproduction of deeply purple spicules.

We could see on the microscope slides that the key agents of gorgonian immunity were the specialized cells called amoebocytes. These cells released antimicrobial chemicals and built walls around infections, isolating fungus from spreading to the rest of the coral. Laura carried out chemical assays showing that the amoebocytes produced an enzyme called prophenoloxidase that initiated a spreading cascade of chemical reactions that included antimicrobial peptides and those that make up melanin. Prophenoloxidase and the cascade of immune activation is a foundational part of the innate immune system and had previously been found in all invertebrate immune systems so far studied (mostly terrestrial insects and worms). Laura had just added corals. What we were seeing inside the sea fans was exactly what would happen inside the much better studied common fruit fly or in a snail when infected with a fungal pathogen. We had discovered that the immune response in one of the most ancient of invertebrates was very similar to that of much later evolved groups on land like insects and snails. Since the basics of immunity are similar in all these invertebrate groups, it implies a common evolutionary origin hundreds of millions of years ago. If this origin was early enough in the evolution of animals, it may have been passed along to form the basis of the ancient arm of the vertebrate immune system called innate immunity.

As we learned more about the gorgonian immune system and gained experience working with a live pathogen to trigger defenses, I began to think about our earlier search for drugs from sponges and the possibility that our sea fan work could be relevant in the search for new drugs from the sea. New chemical technologies were allowing discoveries to come fast and furiously, and all eyes were on marine organisms as a rich source of new drugs. Sponges were still the number one source, but other tropical, sessile animals like corals were yielding treasure troves of chemicals with potential medical value in humans. Our underwater experiments led us to explore more broadly octocorals' capability to kill or slow growth of fungi. If we could find a compound that fought *Aspergillus sydowii* infections, or infections of other pathogenic fungi or bacteria, it could be a lifesaver for the immune-compromised humans most susceptible to such infections.

Over several trips, we collected thirty-six species of Caribbean gorgonian, from sea fans to sea whips to sea plumes in colors from pale purple and bright yellow to black, tan, and brown. We sent them to our chemist collaborator, Dr. Paul Jensen in William Fenical's lab at Scripps, who tested them against a panel of bacteria. Although we detected antimicrobial activity against pathogenic bacteria of danger to humans, there was little antimicrobial activity against certain marine bacteria, and the chemicals unfortunately ended up being either too toxic for all cells or not quite strong enough to be successful drugs. But the work invigorated my growing interest in bioinspired design and the potential for novel drug discoveries in nature.

In 1995, I returned to our marked transects in Akumal, Mexico, for our annual survey of sea fan health. We had moved our surveys of sea fan health from the Florida Keys to the Yucatán Peninsula in search of healthier and more diverse reefs. Akumal was a small, dreamy town south of the crazed frenzy of Cancún, near some of the most diverse shallow and deep reefs in the region. Shallow and deep reefs in the 1990s were dominated by healthy hard corals and gorgonians.

On this day we were headed to a deeper site at a fairly remote deep reef, starting at fifty feet. We motored around in our small boat searching for the exact location. It was always hard to relocate our deeper marked sites from the surface, even using a combination of GPS coordinates, shore markers, and our Mexican boat captains' memories. The faint shape of the darker reef against lighter sand looked right, so we dropped off the boat and descended on scuba slowly through clear, blue water to the reef ridge below us. I was expecting to find many of our marked sea fans dead. What I saw instead made me double-check the location, but we found our old rebar markers, so we were in the right place. Then we saw our large sea fan colonies, a meter tall and marked with numbered metal tags. They were ratty looking, with former lesions detectable as jagged edges, but they were alive! The lesions had filled in and regrown skeleton, and there was no active disease. Among the recovered giant old sea fans were some tiny young ones, brand-new recruits to this reef. It was extraordinary to find that these fans had prevailed against a disease that had killed so many. This pattern of survival, of sea fans fighting back and winning the war against a deadly disease, was replayed on reef after reef across the Caribbean that year. We were seeing, in real time, what scientists call a strong selective pressure. The sea fans with the most robust immune responses had survived and were going on to produce the next generation of more-resistant sea fans. I felt like I was back at the dawn of coral evolution, in the shallow seas of the Silurian period, watching the original arms race between corals and microbes—but with the awareness that the results of that struggle would be parlayed into success for the corals hundreds of millions of years later.

When we talk about biological superpowers—attributes that are so powerful and far-reaching in their consequences that they defy what we think of as normal rules of biology or are so complex that we do not yet understand how they work—we have to consider immune systems. As humans, we are mostly focused on our own immune system when we seek a greater understanding of immunity. That limited scope works reasonably well because immunity operates the same general way in all animals with backbones. Vertebrate immune systems, including that of humans, have two pathways: an adaptive component based on antibodies and also an innate component, perhaps inherited from early invertebrates. The invertebrate immune system has only the innate component. Some scientists view innate immunity as less effective than adaptive immunity, because the adaptive immune system learns to recognize a particular virus or bacterium and build the tools needed to destroy it and then remember that virus (or viral proteins in a vaccine) forever, or at least a long time. But the adaptive immune reaction is slow to develop in response to the initial infection, while innate immunity is almost instantaneous in its defense.

Innate immunity of humans is our first and fastest line of defense against a pathogen. It acts quickly and nonspecifically to stop a microbe until the adaptive arm of our immune system can activate a more specific and longer-lasting protection. The properties of our innate immune system are descended from the same ancient source as the seemingly simpler systems in invertebrates like corals. The human innate immune system shares at least three basic components with the immune system of corals: a toll-like receptor that recognizes foreign bacterial or fungal invaders, large immune cells that engulf an invader, and antimicrobial chemicals that poison an invader. The toll-like receptor is like a watchdog that recognizes and responds to a variety of foreign threats inside our bodies, like bacteria, viruses, and fungi. When triggered by one of these intruders, this watchdog receptor activates a cascade of signals that call other immune cells to engulf the threat and release inflammatory cytokines and antimicrobial chemicals. The human innate immune system is relatively nonspecific and activates against any foreign microbe. This is essentially also how the coral immune system works.

One interesting twist in our human struggle with SARS-CoV-2 is that there was early evidence that innate immunity was important in resisting infection. Because innate immunity wanes with age, I think this may have been why younger people were initially more resistant to COVID-19 than older people. Our scientific establishment initially pursued vaccines to beef up adaptive immunity against COVID-19, but could we have also focused on innate immune processes for more immediate protection? I read with excitement an article that tackled this very question.

In an article entitled "Old Vaccines for New Infections," Konstantin Chumakov (associate director of vaccines at the FDA) and Robert Gallo (director of the Institute of Human Virology at the University of Maryland, and discoverer of HIV) discuss the value of vaccines that activate our innate immune system. One example they cite is the bacille Calmette-Guérin (BCG) vaccine for tuberculosis. It contains a weakened form of a bacterium close to the one that causes tuberculosis. This is a live-attenuated vaccine (LAV), like the ones used to target measles and polio, that works by first inducing early protective innate immunity. LAVs are used for polio because they both activate immediate innate immunity and trigger more long-lasting protection against multiple strains of polio through adaptive immunity. Early in the COVID-19 outbreak, Chumakov and Gallo advocated using existing LAVs (or developing a SARS-CoV-2-specific LAV) to bridge the protection gap until a vaccine that worked with the adaptive immune system could be developed. They argued that the broad innate immune protection induced by LAVs would not be compromised by the evolution of variants resistant to specific vaccines. LAVs might have offered an essential tool to immediately "bend the pandemic curve," but this strategy was not pursued. Even though the adaptive immune vaccines eventually developed in a remarkably short time were a spectacular triumph, I continue to wonder what we might have learned about battling SARS-CoV-2 and other viruses if we had put more credence in the innate arm of our immune system and harnessed our inner coral. It is a reminder that the ocean's most ancient animals still have much to teach us when it comes to battling disease-causing microbes—a game they've been playing far longer than we have.

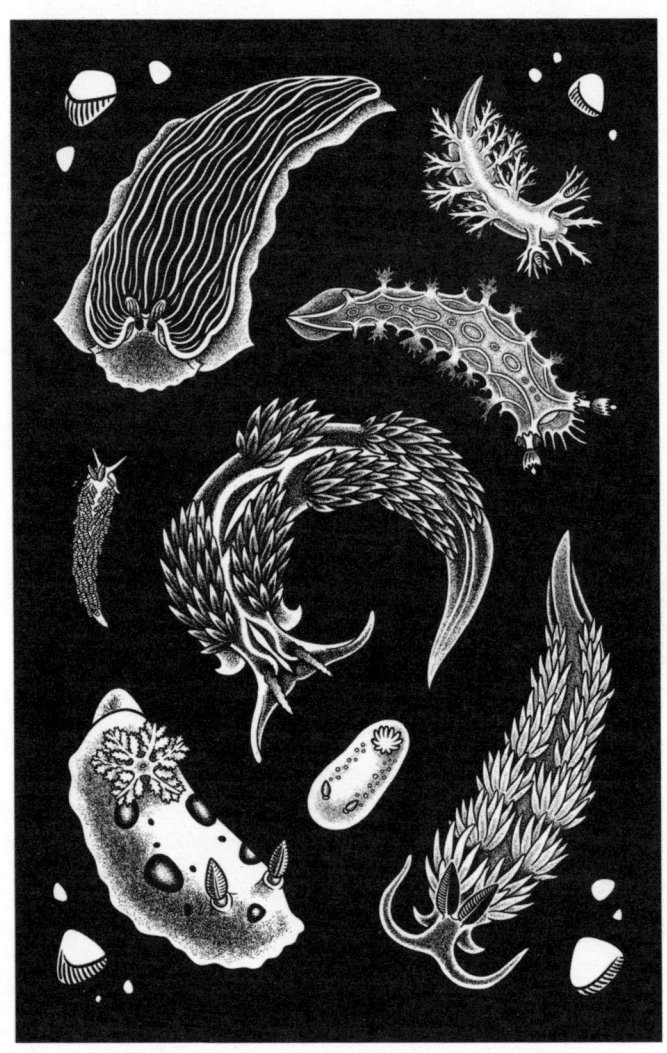

Sea slugs are relatives of shelled snails and come in many forms, including the predatory nudibranchs from four different families (clockwise from top left: two frilly dendronotids, three spiky aeolids, and two dorids) shown here.

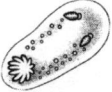

4.

The Sea Slug's Sting

We don't know much about the earliest microbes that stirred in the earth's soupy oceans a billion years ago in the Neoproterozoic era. One thing we can be certain of, however, is that they ate each other. We can also be certain that there was selective pressure to be good at capturing prey and equally strong pressure to avoid becoming prey. Ever since that time, evolution has been largely an arms race between the eaters and the eaten. Would-be prey have developed defenses, from hard shells to sharp projections to toxic secretions, to foil the creatures that might have them for lunch. The predators, in turn, have responded by growing harder teeth, stronger jaws, tougher digestive cavities, and tricks for disabling or tolerating toxic chemicals and dangerous harpoons.

When ecologists like me study relationships among organisms, we have this long-term back-and-forth evolutionary dance in the back of our heads as we decode what's going on. Evolution has produced both exquisitely designed defenses and brilliant methods of getting around those defenses, and both sides are on full display within the snails and their relatives.

The well-known defensive solution of many snails is a hard shell. Snails retreat into their shells when threatened, often by fish or crabs, and sport ornate spines and twists that add protections. Some snails, like cowries that eat toxic prey, have not only a simple smooth shell but also unusual powers to load up on toxic chemicals for defense. The marine slugs, or nudibranchs, lack a shell and load up on other kinds of defense. Some nudibranchs make toxic chemicals of their own. Some steal toxic chemicals from their prey and house the arsenal in their own cells. They are essentially stealing their prey's defensive adaptation and repurposing it for their own use.

One group of nudibranchs, the aeolids, raises this steal-and-repurpose strategy to the level of a superpower. They steal something from their prey that's much trickier to handle than a simple poison and often more effective for defense. These are among the showiest of nudibranchs, with brightly colored, fingerlike projections decorating their backs. All this showy finery hides an exploding defense on a hair trigger, tucked in the projections. The aeolids' strategy is so improbable that it took scientists a long time to figure out what they were doing and an even longer time to understand how they pulled it off. It certainly qualifies as a superpower.

Aeolids eat sessile members of the phylum Cnidaria, including sea anemones, but also corals and more mobile jellyfish. As soft organisms that can't flee, sea anemones would be easy prey were it not for one key attribute—their tentacles are armed with vicious stinging cells called nematocysts or cnidocytes. Nematocysts are so crucial to the survival of sea anemones and other cnidarians, like corals and jellyfish, that scientists consider them a defining characteristic of the whole phylum.

Nematocysts are a wonder of nature. Imagine a sharp-tipped tiny harpoon, connected by a hollow thread to a toxic venom reserve, coiled into a protective capsule with a hair trigger. They fire in less than a millisecond, one of the fastest defenses in nature. The harpoons launch at an ultrafast speed of eighty meters per second, the fastest speed known for a biological response, and quicker than the blink of an eye or a sprinting cheetah. No bigger than fifty microns, the capsule is too small to be seen by the naked eye. The nematocysts in some anemones and jellyfish inject powerful venoms, some strong enough to disable or even kill a human. One of the most venomous nematocysts comes from a box jellyfish. A sting not only hurts like hell, it also causes swelling, painful muscular spasms, skin necrosis, a rapid weak pulse, prostration, lung swelling, heart failure, and breathing depression. It can kill you.

Nematocysts are pretty handy organelles, used by sea anemones and other cnidarians like jellyfish primarily to acquire food. When a fish or snail wanders too close to the waving tentacles, it gets stung multiple times, each nematocyst delivering a charge of toxin. The stunned prey is then swept into the anemone's digestive cavity. But nematocysts work equally well for defense, and the nematocysts make lots of different kinds from sticky threads to penetrating harpoons. For an anemone, it's like having rows and rows of expert archers stationed in the crenellations of its castle, all armed with poison-tipped harpoons. Any creature that tries to take a chunk of flesh gets a mouthful of these nasty harpoons.

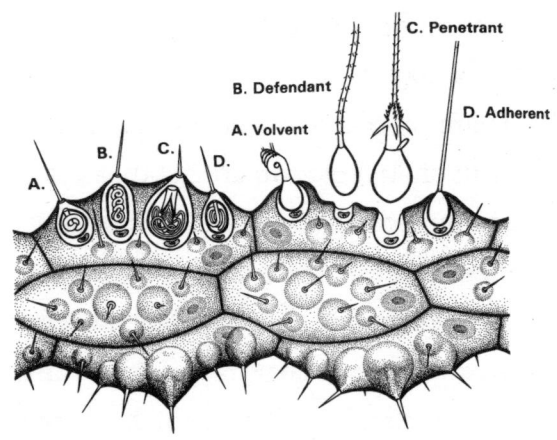

Multiple types of nematocysts made by cnidarians for prey capture and defense.

DIAGRAM COURTESY OF VICKI PEARSE, *ANIMALS WITHOUT BACKBONES* (CHICAGO, IL: THE UNIVERSITY OF CHICAGO PRESS, 1987)

Aeolid nudibranchs specialize in eating only cnidarians. How do the soft-bodied aeolids overcome this formidable defensive adaptation when attacking an anemone?

The tide pools that dot the rocky shores of the Pacific coast are peekaboo worlds of marine life, each housing an ecosystem crammed full of red and green algae and an unpredictable cast of stunning creatures, from five species of slithering snails, to crabs from hermit crabs to rock crabs, to jetting tide pool sculpins, to slow-moving chitons and limpets. Many are vibrant gardens of the green, pink-tipped anemone, *Anthopleura elegantissima*, spreading tentacle to tentacle across the rock. This clonal anemone reproduces by dividing in half. Within a day, an anemone produces an exact copy of itself. This ability allows the anemone to quickly create a meadow of fuchsia-tipped tentacles against a background of green-striped mouths in even the largest tide pool.

Tide pools full of *Anthopleura* are good places to look for one of the more common intertidal nudibranchs, *Aeolidia papillosa*. This off-white nudibranch, often called the shag rug sea slug, is mottled with speckles of gray and covered with plushy projections that make it resemble a fluffy Easter Bunny. But this harmless appearance hides a deadly identity. The shag rug sea slug is a vicious carnivore, one of the few animals capable of eating *Anthopleura elegantissima*.

I've watched shag rug sea slugs take their prey many times. The first time I observed the whole process is still fresh in my memory. I had found a particularly large *Aeolidia*—almost five inches long—at the edge of a large pool on the shores of San Juan Island. I set up my camera and sat beside the tide pool, filming the nudibranch as it patrolled the pool margin. I got a good look as it glided past, noticing dark eye spots and even its heartbeat pulsing through translucent skin as it entered the tide pool and stretched out its white-tipped chemosensory antennae (called rhinophores), slipping silently on a ciliated foot and sniffing with swiveling rhinophores. It detected the anemones and locked on to the source, swinging its rhinophores toward an anemone. Slowly, at its snail's pace, it moved closer and closer to what seemed an unaware anemone, gliding smoothly overtop stationary chitons and limpets along the way.

The anemones are sessile and attached to the rock bottom of the pool, so it was unclear to me why the *Aeolidia* approached in such a careful, almost thoughtful way. It's not as if the anemone would run away. Maybe there are chemical differences between anemones, and the nudibranch was making sure to nab the correct one. Finally, when it was millimeters away, the *Aeolidia* raised its head in slow motion and reared the whole front half of its body off the rock. Then it pounced like a cat, striking hard and fast, gripping implacably with sharp-toothed jaws. The anemone contracted, pulling in its lethal tentacles, and thrashed back and forth to dislodge the killing force, but the nudibranch hung on tenaciously, like a lion taking down an antelope. The anemone gradually stopped thrashing as the *Aeolidia* proceeded to slowly devour it.

Maybe the *Aeolidia* pounced on its victim because that all-at-once attack is what allowed it to avoid the stings of the anemone's nematocysts. *Aeolidia* is a master at producing slime, a mucus thick and sticky enough to contain the nematocysts' harpoons and keep them from penetrating its own skin. But the slime—produced on the sea slug's underside—could do its job only if it covered most of the anemone tentacles all at once, causing the nematocysts to fire harmlessly. A more leisurely attack might expose the sea slug's head and sensory rhinophores to stings. The takedown was a neat trick. But it enabled something even more awe-inspiring. Contained in the anemone flesh the sea slug ingested was not just nutrition but also nematocysts, the key to the sea slug's defensive superpower.

Watching *Aeolidia* hunt its dangerous prey aroused my interest in the broader question of how carnivorous sea slugs like *Aeolidia* and some tropical cowries eat highly defended, dangerous foods. These foods include prey loaded with toxic compounds, prey armed with nematocysts, and prey with both kinds of defenses. This puzzle involved questions not just about the mechanisms of detoxification of poisons and neutralization of nematocysts, but also food choices and coloration. The color palette and patterns of the nudibranchs are showstoppers and are related to their ability to handle explosives and toxic chemicals. Nudibranchs sport every color of the rainbow and often with vivid contrasting stripes or polka dots. Lime green or pink stripes pop against black; yellow stripes on bright blue or orange spots on white or simply hot pink flaps swirl on their backs. Nudibranchs are famous for eating toxic or defended prey like sponges, corals, and anemones, and much of the color show is thought to advertise their high level of defense to visual predators like fish.

With the enthusiasm characteristic of a graduate student, I wanted to know more about all kinds of carnivorous gastropods. Any knowledge I could gather on one species would help me understand them all. Thus, I was beyond excited when I was invited, in 1983, to be part of a weeklong saturation diving mission off Saint Croix in the Virgin Islands, run by the National Oceanic and Atmospheric Administration (NOAA). This first venture as a saturation diving researcher was followed by the chance to do three subsequent missions. As described in chapter 3, in the first two missions, I lived with three other scientists in an underwater habitat called Hydrolab anchored fifty feet below the surface and studied the aggressive interactions between neighboring corals on the tropical reef. Our coral research kept me busy, but I also had time at day's beginning and end to undertake my own study on distant cousins of the aeolids, predatory shelled snails called flamingo tongue cowries. In 1986, the Hydrolab was replaced by a larger, more modern underwater habitat called the Aquarius. As a research professor at Cornell, I would go on to lead my own two saturation diving missions in Aquarius.

R\ight beside the saturation diving habitat was a small but diverse reef. It supported many species of soft corals called gorgonians—the only food of the flamingo tongue cowrie and famously toxic. After years of studying carnivorous nudibranchs in cold temperate waters, I was curious and excited to get a look at a tropical snail I had heard about. The flamingo tongue cowrie is gorgeous; it's like an underwater leopard covered with black-rimmed, bright orange spots, but if you are a soft coral, it is a deadly predator. It was also very common. On my first afternoon survey, I donned gear, including a hookah rig with a tether to air from the habitat. The tether was about one hundred feet long, limiting how far I could work but offering the freedom of breathing underwater and a radio in my face mask, without having to carry a tank on my back. Because I was out on the reef alone, as a safety measure, I would maintain continual communication via radio with both the research team in the nearby habitat and the surface team.

I took an underwater camera with strobe, an underwater writing slate, and some flagging tape to mark gorgonian colonies if I found any being attacked by snails. I swam down from the porch of the Hydrolab to the reef. The water was vodka clear, and even at a depth of sixty feet, I could easily see for seventy feet across the reef.

THE SEA SLUG'S STING

To start, I marveled at this little coral oasis on the bottom, populated with yellow and green brain corals, brown pillar corals, and reddish-maroon mounding corals. I was pleased to see, intermixed with the reef-building hard corals, gently swaying gorgonian soft corals in the full diversity of shapes, from six-foot-tall pale white sea plumes to five-foot-tall purple sea fans to shorter dark brown, purple, and black sea rods. In all, there were about fifteen species of hard coral and at least that many gorgonian coral species just on this patch of reef the size of two football fields.

I swam over the reef, trailing the air hose that connected me to the habitat, and reported by radio to the team inside that all was well on my reef. That was an understatement—I felt like I was living the dream. Instead of watching tiny sea slugs under a microscope in the lab, I was out tracking cowries in the wild, on a coral reef. My only worry was whether my test project would work out. Others had reported seeing a few flamingo tongues on this reef, but would there be enough to do a real scientific study of their foraging? I would need at least twenty snails to make it scientifically valid.

I saw the first cowrie on a sea fan right away because it had eaten away a ten-inch scar of purple tissue, leaving a patch of dark skeleton showing against the lavender mesh of the living sea fan. My cowrie was just sitting there near the scar, still eating away, kind of like a caterpillar on a leaf. Another cowrie was at the base of a long black scar on a white sea plume. Even underwater, they stood out, with bright orange, pink, and yellow spots rimmed with black. The vibrancy comes from the colors on the skin, living tissue that overlies the shell. The shell itself is a pretty cream color. Although their beautiful skin looked bright against the sand, the snails were hard to see when nestled among the polyps of some coral prey. I pondered the mystery of its coloration: Was it for warning or camouflage? Was the snail shouting, "I'm poisonous, don't eat me!" or whispering, "You can't see me because I'm camouflaged"? I never fully cracked the code of their skin color; they looked different depending on their background and age.

Time passed in a blur that afternoon as I searched and found dozens of flamingo tongues around the small reef. I photographed each one and recorded on my underwater slate which gorgonian species it was feeding on. In all I spotted over thirty snails attacking about fifteen different species of soft coral. I was as happy and excited as a kid winning a soccer game—this was enough to do a test study. Now all I had to do was mark each snail and follow its sojourn around the reef.

Flamingo tongues eat only gorgonians, but within the gorgonian group the range of species they eat is surprisingly wide, from the bright purple mesh of sea fans to the multicolored sea rods and sea plumes. In total, they eat most of the thirty-nine species of Caribbean gorgonians. But what we didn't know was whether particular snails had specific prey preferences. Did one snail eat only sea fans while another ate only sea plumes, or did each snail eat many species? Making observations of individuals' foraging behavior was necessary for solving the essential puzzle: How does this fast-moving snail with leopard spots eat dangerously toxic soft corals?

Over the course of the week, I saw that Snail Number 4 moved from a sea fan to a sea plume and then on to a sea rod and finally back to a sea fan. The other snails were similarly busy and on the move. During the week, each snail munched on five to ten different species of soft coral, although there was a clear preference for the sea fan. This behavior needed explanation. It takes both an expenditure of energy and exposure to danger from hungry fish to climb down off one colony, transit around the reef, and climb back up on another. It seemed it would be safer to stay in one place, or at least to minimize travel between gorgonian colonies. I developed the hypothesis that this foraging behavior had something to do with managing the toxicity of the gorgonians.

At the time I did the research, we knew little about the gorgonians' chemical defenses—only that the animals were indeed toxic. We went on to learn in future research and saturation diving missions that some of the gorgonian corals possessed deadly chemicals that were studded with rare and impressively toxic elements. It was as if the corals were shopping from the periodic table, selectively picking the most toxic elements to use for protection. Bromine is corrosive to living tissue when liquid, with irritating vapors that can cause dangerous brain changes in humans. Chlorine is even more toxic. The chlorinated and brominated chemicals present in gorgonians would later cause skin swellings and breathing problems for some of my graduate students when studying them in the lab. We learned that flamingo tongue cowries and many other gastropods were chemists extraordinaire, not only tolerating these poisons but also storing them in their skin to be used for their own defense.

How animals manage to eat and then reuse toxic prey is a long-standing question in biology. Among the first scientists to tackle the question was Ernst Stahl, who studied herbivorous land snails in the late 1880s. Like me, he wondered why some snails ate diverse plants and others only a single species. Stahl's motto, "My laboratory is nature," resonates with me to this day. He studied the foraging behaviors and plant preferences of a group of land slugs, including *Helix* and *Limax* species. He was also an experimentalist, intensely interested in how the plant chemicals defended the plants. When he found a plant that his slugs would not eat, he removed the chemicals by extraction and was then able to show that the slugs would eat the chemically undefended plant. He also observed that his snails were divided into species that ate a whole range of plants and specialists that would eat only a single species of plant. Stahl's work is now regarded as a foundation of chemical ecology, which is the study of the chemically mediated interactions between organisms and their environments.

In part because of its many applications, a rich area of chemical ecology research is the biology of caterpillars and their chemically defended host plants. Many terrestrial caterpillars eat toxic plants and sequester their chemicals for defense, just like many marine snails. Consider the bright orange monarch butterfly. Its caterpillar larvae eat milkweed, which is packed with toxic chemicals. As a growing caterpillar eats a whole plant full of toxins, it sequesters in its body a group of toxic chemicals called cardiac glycosides. These heart poisons can stop birds in their tracks and even kill ones that get a big dose. More relevant for understanding marine gastropods is the many caterpillars that eat a diversity of toxic plants. Scientists suggest these caterpillars are eating a diversity of plants to dilute or even detoxify the chemicals in each. This hypothesis of toxin dilution is the explanation for why some caterpillars mix their diets by chomping on multiple species instead of eating only one toxic plant. The toxin dilution hypothesis posits that the mixing of diverse chemicals prevents the most potent ones from exceeding a threshold level beyond which they become dangerous. A possible synergistic effect may also be that some of the mixed-in chemicals detoxify others.

Reviewing this literature, I surmised that the diet mixing I observed among flamingo tongue cowries was a clue to some kind of detoxification strategy. Was there one very nutritious species, like the sea fan they seemed to prefer, that might be too toxic to eat full time? Could they be using the gorgonians' chemicals in their own defense or putting special chemicals in their eggs to protect them? There was no time to do more on this trip, but I was to return a few years later to lead two saturation diving missions and carry on the work with flamingo tongues and gorgonians. After the close of this first weeklong mission in tropical waters, I returned to the cold waters of Puget Sound and my dissertation research with carnivorous nudibranchs, many questions rolling around in my mind.

I had gained a deep appreciation for what carnivorous gastropods are able to do. The cowries' ability to handle toxins was certainly impressive, but even against this standard, the aeolid nudibranchs that lived around Puget Sound were performing a feat that defied biological explanation.

The plushy projections, or cerata, on the backs of aeolids puzzled scientists for a long time. Their bright colors and beauty called for appreciation, but the way attacking fish recoiled suggested they had a bigger purpose. In 1843, the famous systematists Joshua Alder and Albany Hancock imaged with their microscopes tiny, hairlike threads projecting from cells in cerata of *Aeolidia*. They could only speculate on their origin and purpose. Less than twenty years later, the physician T. Strethill Wright nailed part of the answer. He surmised that the tiny hairs in *Aeolidia* were the threads of fired nematocysts. Because nematocysts were known only in cnidarians, Wright proposed the crazy-sounding hypothesis that they were somehow derived from the anemones the nudibranchs were eating. Everyone agreed the threads were from nematocysts, but other scientists thought it more reasonable to suppose that the nudibranchs made the nematocysts themselves. Two other prominent biologists, Julian Huxley and Philip Henry Gosse, agreed with Wright that the nematocysts were made outside the nudibranchs.

It turns out Wright was hot on the trail of the truth. I can relate to the fervor that drove him to design experiments to solve this puzzle. He was asking a specific question about the origins of those nematocysts in the nudibranchs but sensed that he was exploring a fundamental question in biology: the ability of a body to recognize self from nonself and the limits of immune systems in defining species boundaries. If the nudibranchs did not make their nematocysts, how on earth were they acquiring them, and why did their immune system allow it? I will bet that Wright, even in 1863, suspected a major rule in biology was being bent or broken. He took the next step to bolster his hypothesis by discovering that nudibranchs had varying nematocyst types that matched the same types present in their anemone prey. He proved the case with a rather simple but definitive experiment. He fed a nudibranch a new species of anemone, and it acquired nematocysts that matched those of the new anemone. But even so, for the next thirty years, scientists ignored his arguments and clever experiments, incorrectly reasoning that nudibranchs manufactured their own nematocysts. They could not see beyond what they thought were the "normal" rules of nature: cells and tissues were not shared between different species and certainly not between major groups as different as mollusks and cnidarians. A case of one species acquiring and reusing an organ from a different species would have seemed impossible, and frankly, it is still very unusual in our current understanding of the biology of immune systems.

Finally, in 1903, Oxford lecturer George Grosvenor designed experiments that definitively established that the nudibranch nematocysts were stolen from anemones. His persuasive observation was that different nudibranchs of the same species had varying nematocyst types that matched special types from their chosen anemone. Even more impressively, he observed that a single nudibranch could have multiple sizes and shapes of nematocysts when it fed on diverse anemones. The coup de grâce: when a nudibranch was not feeding on anemones, it had no nematocysts.

Although the overwhelming evidence convinced scientists that the nudibranchs were acquiring nematocysts from their anemones, no one knew how the trick was accomplished. The harpoons inside nematocysts are on a hair trigger and fire if jostled or otherwise activated. How could they be transported within a nudibranch's body without firing? Grosvenor suggested that nudibranchs wrapped the nematocysts in a mucous shield, and this allowed them to ingest nematocysts cushioned against firing. Was the answer this simple?

Did the nudibranchs wrap these harpoons in mucus to cushion them against firing and protect them as they pass through the mouth, down the pharynx, through the stomach to their final destination in the projections on their backs? Dr. Otto Glaser, University of Michigan, wrote an entire book about this enigmatic process in 1910 and correctly identified special cells called cnidophages that package and transport the nematocysts.

In 1926, Andre Naville, Faculté des Sciences , Université de Genève, suggested a new idea. He proposed that nudibranchs somehow selected immature, inactive nematocysts. These inactive nematocysts, still encased in their nematocyte cell, could survive the journey through the digestive system and into special cells in the cerata without firing. The idea was so biologically novel and ahead of its time that it sat largely untested for sixty-four years. Then, in 1984, Paul Greenwood and Richard Mariscal thought of a way forward by counting the ages and types of nematocysts ingested as nudibranchs were feeding and digested after feeding. They detected a high proportion of immature nematocysts being delivered to the cerata and a lot of discharged nematocysts in the pharynx and stomach. Then they also discovered that the young nematocysts in nudibranchs were housed in special places in the basal cnidosac, behind the ready-to-fire mature ones.

More recently, new studies have unveiled the way special cells called cnidophages encapsulate and transport the immature nematocysts to deployment sites in the cerata. To this day, the details of how nudibranchs select immature nematocysts, bundle them, and get them to the cerata are unknown. The exact mechanism likely even varies in different species.

Yet another enigma is how the immature nematocysts continue developing inside aeolids' cerata to the point they eventually become working weapons. It is not just a simple parlor trick, but a momentous accomplishment that rewrites a basic biological rule—that nonself cells cannot exist, let alone develop, in an organism's body. The "destroy nonself" rule is a hygiene issue, meant to guard against foreign invaders like pathogenic bacteria, fungi, viruses, and cells of other species. This case of an entire organelle being deliberately stolen from animals in another phylum, tolerated in its new host, and then completing its maturation and function in the body of another species is an extreme rewriting of this rule. The plot thickens as we realize that this is not the first time this has happened.

Nudibranchs are not the only ones to take advantage of stolen nematocysts—a process scientists now call kleptocnidae, aka "stolen cnidae" (an alternative name for nematocyst is cnidocyst). As we know today, kleptocnidy has evolved at least four times in the animal kingdom. Ctenophores, flatworms, acoel flatworms, and nudibranchs all steal and repurpose nematocysts. This amazing trick may be more broadly distributed in nature, but it is best understood in nudibranchs. Within the nudibranchs, it happens in six hundred species, but all of them are in only one group: the aeolids. An ancestral aeolid developed this superpower deep in evolutionary time and passed on variants to diversify in many descendants. That said, there are related groups of sea slugs that do similar things. The herbivorous sea slugs, called saccoglossans, steal photosynthesizing chloroplasts from the algal plants they eat and activate them for their own photosynthesis. As an aside, this reuse of a stolen plant organelle, the chloroplast, is very different from the coral symbiosis with initially free-living algal cells.

Against the 180-year backdrop of this mystery is current work with modern tools of biology to shine a light on the biological mechanisms of how nudibranchs pull off this caper. Scientists have developed tools to study the detailed process in a model nudibranch, *Berghia stephanieae*. It can be grown in the lab and is entirely transparent as a young animal, allowing advanced imaging techniques. *Berghia* have evolved two kinds of special cells to work the magic of selecting and storing nematocysts in their cerata. And of course, it seems like too much, but each special cell has a special name, and here they are! Cnidophages are special transport cells that uptake and transport the immature nematocysts, inside their nematocytes, through the digestive system to cnidosacs, which are special storage cells located in the projections on their backs, called cerata. Inside each cnidosac, there are multiple and sometimes different kinds of nematocysts packaged in their cnidophages. To watch the entire process in these transparent nudibranchs, Jessica Goodheart, an assistant curator at the American Museum of Natural History, feeds juvenile *Berghia* their first meal. Feeding triggers development in the nudibranch of the cnidophages, cnidosacs, and, a bit later, the cerata projections, which will house the arsenal. Nematocysts are loaded into the transport cnidosacs two to four days after feeding. One of the million-dollar questions is where and how selectivity occurs. In *Berghia*, selectivity for nematocyst size and type is very specific. For example, even though there are several types of nematocysts in their prey, they only uptake the most dangerous ones. Selectivity likely happens inside the cnidosacs and likely also inside the cnidophages, which transport the cnidosacs. The tricky bit is we just do not know how the cnidophage selects. This is the crux of the immune recognition process to be solved in future studies.

Cornell professor Dr. Leslie Babonis creates designer anemones to pinpoint fundamental questions in nematocyst development and evolution. One project is to track nematocyst cells after they leave the anemone and enter the nudibranch. A transgenic animal is one with new genes introduced from another animal, like a dog with the genes for a pig heart. Transgenic anemones are created by the CRISPR/Cas9 editing process, which splices a bit of the gene for GFP (green fluorescent protein, originally from cnidarians) into the anemone's genome. The GFP then flashes a fluorescent light anywhere nematocysts are developing. Babonis's dream for future research is to feed a transgenic anemone to the nudibranch and watch in flashing lights as immature, labeled nematocysts march down their normal developmental pathway in the wrong animal (i.e., a nudibranch instead of an anemone).

Kleptocnidae is a defining character of the aeolid nudibranchs even though as a group these six hundred species eat diverse cnidarian prey with different types of nematocysts that vary in their effectiveness. The stolen nematocysts can sting fish and shrimp predators and sometimes the stolen nematocysts even sting humans. The blue dragon is a small, planktonic nudibranch that feeds on the jelly of the heavily armed Portuguese man-of-war. Australian surfers have been stung by the potent nematocysts from the blue dragon.

Nudibranch kleptocnidae is a superpower that bends two biological rules, and this could matter to us humans. The first superpower lies in the ability of the nudibranch to provide a developmental environment that activates and supports maturation of the anemone's nematocysts. One recent study shows with pH-sensitive dyes that the cnidosac cells can tune their internal pH level and that a more acidic environment initiates maturation of the nematocysts.

Secondly, the nudibranch's immune system is strangely permissive to this foreign body, the nematocyst. Immune systems are normally very specific in recognizing cells that come from the outside and, in particular, those that come from unrelated organisms. For example, if you wanted a kidney transplant from a cow or a pig (mammals in the same class as humans), you would need to take powerful drugs to suppress your immune system so that it would not destroy the new kidney. An anemone is on the far side of the tree of life from a nudibranch and is definitely an immunologically alien animal. Anemone body parts traveling through the nudibranch system should trigger big alarms. One factor that helps this feat be accomplished is that the nematocysts travel in the digestive system. Digestive systems do generally have a different set of immune rules, as they are used to handling foreign tissues and are generally more permissive than other body parts.

Overwriting these two biological rules allowed our shag rug nudibranch, along with other aeolids, to deploy fully weaponized cerata filled with hundreds of anemone nematocysts, all ready to sting a fish unfortunate enough to get too curious or too hungry and brush against its back.

We still do not know exactly how nudibranchs successfully transplant and subsequently control an organ from a different species into their own bodies, but three possibilities come to mind. Nudibranchs might prevent immune rejection of nematocysts by inhibiting the recognition of them as foreign, by preventing the activation of immune cells, or by suppressing the production of inflammatory cytokines. It is also possible that the nematocysts themselves have special properties to avoid detection by the immune system. For example, cnidarians produce a range of toxins and other molecules that can affect immune function, and it is possible that some of these molecules could play a role in suppressing the immune response of nudibranchs that acquire their nematocysts.

Understanding the biological controls on several steps in this immune-mediated takeover by nudibranchs could inspire new approaches to transplantation of organs into humans. Transplanting any different organ even from one human to another, whether it is heart or liver or kidney, involves its own special challenges. The example of kidney transplants is easiest to talk about: kidney transplants are more successful than other organs because kidneys are more immune tolerant than other organs. Organs such as the pancreas or heart are much trickier to transplant. However, even if the rejection of an organ from a different person could be medically solved, all human organs are only very rarely donated. So looking ahead to the biomedical need for healthy organs raises the appeal of xenotransplantation—the use of organs from different species. Pigs are a species considered a promising donor for humans because their organs are a good match in size and function, and their immune properties can be engineered to reduce immune rejection. Researchers have successfully transplanted pig islet cells into humans, where they manufacture insulin as a treatment for type 1 diabetes. Active research is focusing on the bigger project of how to transplant pig kidneys to humans, because this is judged as the most likely organ transplant to succeed due to the immunological permissiveness of kidneys. This brings us back to nudibranchs and what they have innovated. Nudibranch kleptocnidae is basically a xenotransplantation, so understanding how nudibranchs handle the immunological challenge of using organelles from another species could help expand our ability in human xenotransplantation. If we think of an immune response as a two-step process that involves first recognition of nonself and then rejection of nonself, the rules bent by nudibranchs appear to be in the recognition process. The nudibranch cnidophages actively select and carry the nematocysts, so some recognition signal must be deactivated in the process of selecting the nematocysts. The usual medical approach to transplanting organs in humans is to suppress the immune system, but thinking about nudibranch tricks makes me wonder if there is a more specific strategy that reaches earlier in the process. Right at the start, could we train the human immune system to code transplanted organs as self rather than nonself and avoid the entire immune rejection chemical cascade?

While it seems like a big stretch to use an invertebrate process such as kleptocnidae as the model for human organ transplants, it is not any more far-fetched than some of the ways other invertebrate superpowers have inspired radical applications to human needs. In chapter 7, you will read about how a fluorescent protein discovered in jellyfish became a mainstay of biomedical research and won a Nobel Prize, and how the skin of octopuses and sea stars is aiding new approaches to development of artificial skin and limbs. The wonder of many of these invertebrate superpowers is the sophistication of biological processes innovated to produce them and the opportunities they provide us to see a new approach to old problems.

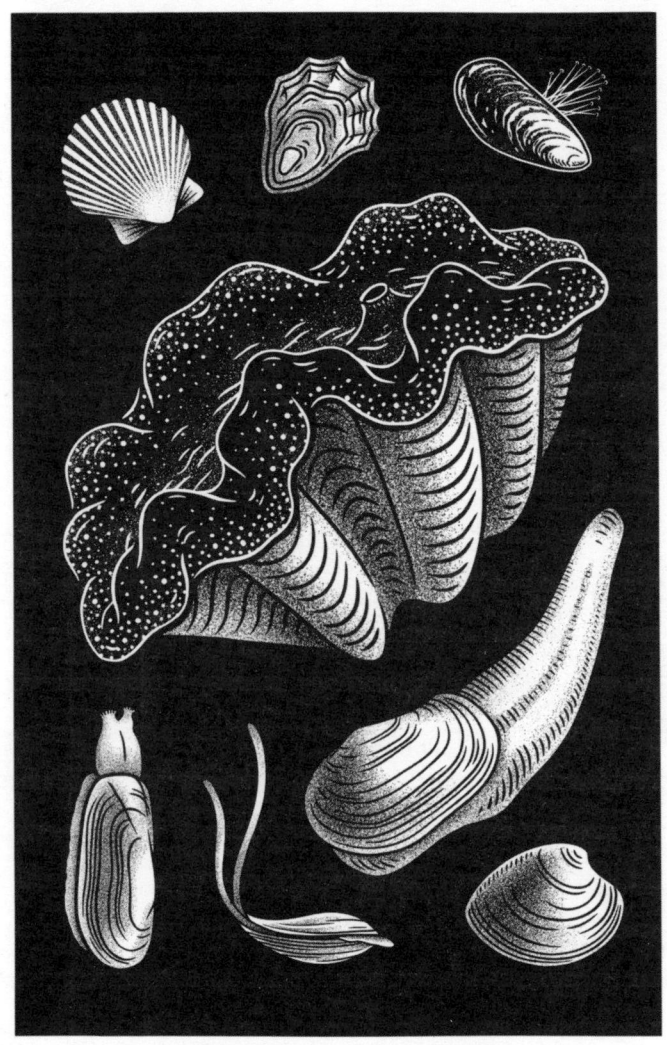

The diversity of animals we call bivalves, from the burrowing clams that live deep in sand or mud (shown lower in the figure) to giant clams that perch on a reef to a big mix of oysters, clams, and mussels that attach to rocks.

5.

The Giant Clam's Light Trick

My introduction to giant clams came as an accident. I had seen these harmless monsters during research dives in Indonesia and Palau, but I'd never thought through how they actually work as photosynthesizing animals and the extent of the engineering innovation in how they arrange their solar arrays and adjust the wavelengths of light. That all changed in 2007 with a visit to Bolinao, on a remote peninsula in the Philippines, north of Manila. Fish farms in the sheltered channel between Bolinao and Santiago Island were fueling algal blooms and bacterial growth, mucking up the water and killing the coral. The channel was virtually choked closed by intense fish farming. Overfeeding with food pellets was creating nutrient-filled, murky water that was harming the nearby reef. Our job was to figure out whether the many aquaculture farms were not only exporting too many nutrients in the water but, maybe worse, exporting killer bacteria that was sickening coral.

We could see from the murky plumes that the fish farms were creating poor water conditions, and we would use new sequencing approaches developed in the lab of our microbiologist partner, Professor Farooq Azam at Scripps Institution of Oceanography, to drill down to quantify the levels and identify the actual kinds of bacteria that could be killing corals near the fish farms. For starters, the corals near the fish farms had already died, leaving only the old, dead skeletal reef as a silent testimony to some prior disaster. So our work was to re-create a reef by transplanting living coral fragments at different distances from the farms and identify the bacteria that colonized them. Along the way, we spent time underwater with the long-lived, colorful giant clams and learned how conservationists farm them.

Our work, funded by the World Bank and Global Environmental Facility, was to create or support centers of research excellence, and one such center was the Bolinao Marine Laboratory. A big component of support for research hubs is to contribute to global training. An effective way to do this is through a form of near-peer mentoring—having near peers, in this case graduate students, learn from one another and also interact with leaders in their field. I brought along two PhD students from my Cornell research group, Courtney Couch and Morgan Mouchka, as experts in molecular biology and also to allow them the opportunity to gain their own international research experience.

In the course of our work with corals, the researchers at Bolinao Marine Laboratory showed us around their long-running effort to

THE GIANT CLAM'S LIGHT TRICK

restock the Philippines' reefs with giant clams. The foundation of this initiative—a giant clam nursery located on a shallow reef on the exposed, ocean side of Santiago Island—was adjacent to our research sites, and we were brought along on a dive for a firsthand look. We jumped at the opportunity to learn about a new superpower of invertebrates and dive in the clam nursery.

This coral reef, alive with color and light, was very shallow, only fifteen feet to the sandy bottom dotted with coral colonies, sponges, sea urchins, and vibrantly colored giant clams. Small fish darted in many shapes and hues; the largest ones are long since overfished in this region of the Philippines. Although the scattered coral heads were familiar, this reef was different from any I'd seen before. Large plant-eating fish like tangs and parrotfish had been fished out and were replaced by battalions of long-spined black urchins patrolling the bottom, looking like a horde of giant black spiders. Urchins are also plant eaters, and they are actively cultivated in the laboratory and then outplanted on this overfished reef as lawn mowers. They graze back the marine algae so it doesn't shade the corals and young clams, which need light. The reef is highly managed, cultivated as a giant clam nursery covering an area of 5.8 hectares, about the size of seven soccer fields. Healthy corals were intermixed with over twenty-five thousand giant clams in different sizes, from two inches to as large as a small pony at four feet across. The clams in this nursery, including the four-foot-wide giants, were raised in a seawater nursery at the nearby Bolinao Marine Laboratory and then outplanted to reach full size on this protected reef to reestablish their populations. Because they are delicious to eat, can't move, and are very slow-growing, giant clams have been overharvested to endangerment and need special protection like this reserve.

We swam as a threesome in warm, clear water, completely captivated by the variety of giant clam sizes, shapes, and colors. It was breathtaking to see so many of these strange clams, which don't live in the Caribbean and are rare now on most Pacific coral reefs. A large shape like a boulder or coral head loomed ahead, but as we got closer, the characteristic jagged shell was clear; it was one of the giants at three feet high and almost four feet across. The shell was gaped a foot open, so I could see the surface of the clam's skin surrounding its incurrent and excurrent tubes, called siphons, and reflecting iridescent teal and purple splotches. I stopped and breathed, suspended above the clam to watch as it drew water inside through an inch-wide incurrent siphon to filter across its huge internal gill. I couldn't see the gill, but I could see tiny particles swirling into the clam, entrained in the strong current it produced. I was mesmerized by the steady intake of water into the round, blue-rimmed intake siphon and the constant adjustments in diameter to regulate flow as attentively as a mechanical regulator. After it's filtered through the giant gills that fill the inside of the clam and extract all phytoplankton, the water pulses through the outflowing (excurrent) siphon. At the same time, the clam's skin rippled and undulated, and the teal spots flashed as the clam shifted position and regulated flow over its gills. Both incurrent and excurrent siphons were richly decorated with patterns of neon dots. Some of the colors themselves spark energy from the sun; the rich tapestry of spots, splotches, curlicues, and dashed lines is both dazzling and mysterious. Along with the blotches of photosynthetic

THE GIANT CLAM'S LIGHT TRICK

pigments were hundreds of eyespots—round dark spots the size of a ladybug, surrounded by teal outlines—scattered along the edge of the gape. The clam eyespots can't form images and see us, but the clam did detect our shadow and abruptly stopped filtering and warily narrowed its gape. This was *Tridacna gigas*, the largest clam species, which can weigh over five hundred pounds and live a hundred years. We left it in peace and scanned for others. The reef bottom was dotted with a multitude of smaller clams. They were embedded and harder to see, but their bright skin showed as a stunning azure blue or pale green or maroon, dotted with darker eyespots outlining the edges of the jagged shell.

Giant clams are their own taxonomic family, but they are close relatives of more normal-sized clams. The more familiar clams that we eat—like the two-to-three-inch steamer clams or bent-nose clams—live hidden lives, burrowed into sediments, and feed by extracting phytoplankton from water drawn in from a siphon, down a burrow, and filtered over the gills. These bivalves rarely see the light of day, because only their siphon breaks the sand's surface. Most clams are smaller than a tennis ball and rarely open their valves wider than an inch, and then only to extend their siphon. One exception is the geoduck of the Pacific Northwest, which can get larger than a cat, weigh up to fifteen pounds, and chill deep in a burrow with a three-foot siphon reaching the surface.

There are twelve giant clam species in the Coral Triangle, a swath of the western Pacific rich in marine life, and eight can be found in the Philippines. Some of them, like the long-lived *T. gigas* we saw, reach five feet in diameter and weigh five hundred pounds. Among the invertebrates, they are exceeded in size only by giant squid. The key to their large size is an ability to photosynthesize and derive energy from the sun. Although they are truly docile, gentle giants that simply sit on the ocean floor filtering water, their large stature can feel ominous. It's easy to see why fabulous tales once spread that divers were in danger of getting trapped in a clamped-shut shell. The navy's diving manuals once included instructions for how to escape if clamped in a giant clam. This can never have happened. Indeed, the gigantic clam species cannot close its shell completely, let alone snap it shut.

As we toured the giant clam laboratory, we saw a snapshot of the life story of how a giant clam makes its way in the sea from tiny mobile larva to cemented-in-place gargantuan adult. The life story is as improbable as it is epic and also took me back to my own early work with larvae of sea slugs. To start with, keep in mind that on a coral reef, adult clams are anchored in place. Immobility could create issues of proximity in getting one clam's sperm to find another clam's eggs. Problem solved, because bivalves are hermaphrodites; each clam is both sexes and can share either sperm or eggs with any nearby clam.

These clams have other tricks to help them make babies, such as awaiting special evenings, signaled by the phase of the moon and tides, for sex. When there is either a new or full moon and the tide is coming in, one clam starts it off by releasing sperm. Nearby adults detect the sperm and release either eggs or sperm, triggering yet more animals to release a virtual orgy of sperm and eggs in the

THE GIANT CLAM'S LIGHT TRICK

water. The lunar-phased group spawning ensures cross-fertilization of eggs that will then develop into a flotilla of tiny plankton-living larvae.

Two days after the first hurdle of fertilizing an egg, the tiny developing embryo, ten times smaller than a ladybug, develops beating cilia and hatches into a swimming larva. This larva is also tiny, less than one quarter of a millimeter long, with a giant-size job to do in its short life in a wide ocean. While washed around by the currents, it swims quickly and feeds busily on phytoplankton for a week, all the while evading predators. Its nervous system has developed new sensory capabilities, and after a week, it puts these senses to work in its next big job: picking the right spot to spend the next hundred years. As the tiny larva swims, it begins the search for others of its kind, sniffing for chemical cues. The larva needs to find the others because it will not be able to eventually reproduce in three years unless there are other clams nearby.

Once the larval clam picks its home for the rest of its long life, it metamorphoses into an adorable baby clam, still tiny at less than a millimeter, but with the feeding gills of an adult and a small jagged shell. All clams are herbivores and filter tiny plant cells from the water in a suction current created with their ciliated gills. Life is still very perilous for the newly metamorphosed juveniles, which die by the millions. One threat is a cadre of predators, including small snails that hunt down the young clams. *Cymatium* snails are tiny parasites that can enter a clam at the size of one to two millimeters and eat it from the inside out, by injecting a toxin that dissolves the flesh. Despite the dangers to the larval and early juvenile stages, some clams make it through and grow for hundreds of years.

Giant clams are different from all other clams in the great size and age they will reach and the fact that they spend a hundred-year lifetime cemented in place. They are also different from most other clams in having a superpower. Giant clams, like reef-building corals, live in symbiosis with unicellular algae, which turn sunlight into food. They activate this power after they cement themselves into the reef and use their gills to capture the special species of symbiotic algae from the plankton, which go on to divide and populate the entire clam. Then these clams launch a dazzling new life.

Though it would seem to leave them vulnerable, there is a good reason giant clams risk leaving their shells gaped open. Although a giant clam feeds both by filtering plankton from the water and by the solar contribution from its algal partner, the gains from photosynthesis outweigh those from plankton. The amount of solar gain varies by species. *Tridacna gigas*, is the best studied and the most photosynthetically successful. Its reliance on energy from the sun increases as it grows, until a fully grown *T. gigas* clam relies 100 percent on photosynthate from its algal partner, with energy to spare. Their strange open-mouthed stance is so fixed by evolution to expose the photosynthesizing skin that some species can no longer close their shells completely. The iridescent colors in a clam's skin are not only out of this world but vital for its life. The bright colors in the mantle activate photosynthesis, but the real mystery is why these neon colors splash across the skin in such entrancing patterns and how they change the wavelengths of sunlight.

THE GIANT CLAM'S LIGHT TRICK

Giant clams fill special pouches in their skin with unicellular algae that are close relatives of the zooxanthellae that power solar gains in corals. While both giant clams and corals create and use energy from their photosynthesizing algal partners, clams have developed some completely different, whiz-bang approaches to optimizing gains from the sun. Giant clams expanded the superpower of their solar-powered symbionts to the level of a solar reactor by developing a special light-optimizing cell and structural designs to go along with it. While corals also have evolved special features like a partnership with algae and light-reflecting patterns in their skeleton, as discussed in chapter 2, they do not have a specialized cell like the clam's iridocyte. This cell has its own way of changing biological rules by actually bending the light to change its wavelength.

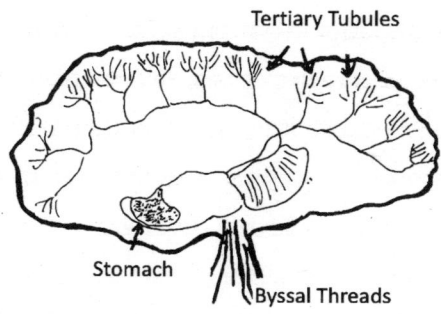

Distribution of the columns of tubules, branched off the stomach, that house the symbiotic algae cells.

DIAGRAM ADAPTED FROM REF 10, "POUCHES FROM THE CLAM'S STOMACH," JOHN H. NORTON, MALCOM A. SHEPHERD, HELEN M. LONG, AND WILLIAM K. FITT, "THE ZOOXANTHELLAL TUBULAR SYSTEM IN THE GIANT CLAM," *BIOLOGICAL BULLETIN* 183 (DECEMBER 1992): 503–6

In giant clams, the surface of the skin is like a disco ball, covered with a layer of sparkly, colorful iridocytes. The iridocytes allow clams to manipulate light via layers of micro-reflectors within each cell. These cells scatter and reflect light to increase the photosynthetic efficiency of the algae and create distinct skin colors.

Under the colorful, sparkling iridocytes lies the real powerhouse: high densities of photosynthesizing zooxanthellae algal cells. Just like the coral algal symbionts, the algae are brownish spheres, approximately ten microns in diameter. These cells are arranged in vertical columns in the clam's skin that stretch toward the inside of the clam, parallel to the direction of incoming solar light. Interestingly, the columns containing the algal cells are thin membranous tubules that develop as pouches from the clam's stomach. Again and again, we see that inclusions of foreign bodies, whether they are algal cells in corals or clams or nematocysts in nudibranchs, are packaged in outpocketings from gut tissue. I think this is because the gut is the site in all creatures that experiences the most novelty in foreign material, because it is common for carnivores and herbivores alike to consume prey completely unrelated to them without triggering immune system alarms. As an aside, this is also true for humans who successfully eat diverse animals from bivalve clams to shrimp to cattle. It's only occasionally that some digestive systems become sensitized and their immune system becomes activated to reject foreign organisms. So storage organs related to gut tissue are a logical nursery or repository for these innovations or partnerships.

Algae and the iridocytes in the clam mantle together constitute a mini–solar reactor consisting of two types of cells, one reflecting and bending light (iridocytes) and the other strongly absorbing

THE GIANT CLAM'S LIGHT TRICK

light and photosynthesizing (algae). Organizing algae into tiny pillars within the skin of the clam increases the surface area of algae available for photosynthesis and allows light to penetrate deeper into the clam. This design—packing energy-producing cells in columns activated by reflective cells—creates an extraordinary micro-solar reactor.

The key superpower that is unique to these clams is the way the iridocytes optimize the spectrum of light to wavelengths best for photosynthesis. Although tiny, the iridocytes are packed with layers to bend light. They contain alternating layers of high-refractive index guanine crystals organized into palettes through which light moves slowly and low-refractive index cytoplasm through which light moves quickly. Most remarkably to me, the palette cell tunes its effect on light depending on conditions, by compressing and relaxing these layers. As a result, "the guanine palettes not only reflect harmful UV radiation but also absorb it and emit light at longer wavelengths that are safe and useful for photosynthesis," Ram Chandra Subedi, a researcher at King Abdullah University of Science and Technology, explains. This increases the amount of photosynthetically active radiation available to the algal symbiont. In this way, both the clam and its algal symbionts are sheltered from potentially damaging UV radiation. As an aside, some of the new mineral-based human sunscreens act a bit like iridocytes by changing the sun's rays to longer, less damaging wavelengths.

Animals on land and under the sea, such as clams, spiders, lizards, frogs, and fish, build a remarkable variety of light-reflecting structures by varying the size, type, and arrangement of guanine crystals. These reflectors are often used for communication or camouflage by scattering or mirroring light or even images. Giant clams use their iridocytes to build a solar reactor. Baby fringe-fingered lizards flash a bright blue tail colored by immature guanine crystals. The mature crystals in the adult cast a different hue and aid in camouflage. Glass frogs reflect light and make their blood cells less visible with guanine crystals. Thus, the biological trick in clam iridocytes to bend the wavelengths of light begins with the ability to produce guanine crystals. Making guanine crystals is a "power" separately evolved in different animals and spread across the tree of life. The versatility and complexity of guanine crystals exceed any artificial materials currently being produced—another win for a bioinspired design—and engineers have been intensely interested in these new leads from nature. But only the giant clams use their guanine crystals to manipulate and optimize light in what is effectively a solar reactor.

Engineers have mimicked the way giant clams use the sun to develop new approaches to generating alternative energy more efficiently, such as in solar cells and bioreactors. One enterprising researcher I met from Yale University was uncovering the mystery of how giant clams arrange the vertical columns of their microalgae to optimize the incoming light. "Evolution is so much more clever than human engineers," Dr. Alison Sweeney says in *The Atlantic*. For instance, traditional solar cells work well in direct sunlight, but not when they get too hot. With the clam's design, a reflective sheen could help solar cells stay cool even when they're exposed to intense sunlight, Sweeney says.

Sweeney's research program is built on an interest in bioinspired design and her insight that the best blueprints for innovation already exist in nature. Her lab focuses on the evolution of biological soft matter and the mechanisms by which they assembled themselves over time. Understanding those mechanisms, she says, may offer the means for creating new biofuels, chemicals, and materials that help sustain planet Earth. She claims that giant clams are the most efficient solar energy system on earth. "By that I mean giant clams take in the greatest fraction of sunlight and convert it into chemical energy. We can compare it to any other system, from tropical rainforests to cornfields in Iowa. Clams . . . have these vertical pillars and a mechanism to redistribute and wrap light around them." The efficiency of the cells and the three-dimensionality of the pillars suggest new approaches for more efficient, damage-resistant solar cells and more spatially efficient solar production of algal biofuels, foods, and chemicals. Photobioreactors for algal products and fuels, and plastic solar cells, could benefit from attempts at direct mimicry of the clam system.

As we swam over the array of bright clams in a kaleidoscope of color from teal blue to bright green to reddish brown, I wondered whether each clam had its own specific color and pattern or whether they could shift color with age or conditions. The broad palette of bright colors from several shades of blue to green and violet is produced by the reflectance between guanine crystals within the iridocytes and chlorophyll within the symbiotic algae. The iridescent blue color is not from a blue pigment—which is rare in nature—but, like the blue in a butterfly wing, it is produced by reflection and interference. Chlorophyll a is the pigment that does the work of photosynthesis. Its levels in the algae also control the wavelength of emitted light and thus the color of the clam's spots.

The colors of giant clams vary depending on region. At our dive site in Bolinao, in the northern Philippines, they're mostly brown with green edges; in the prime diving site of Anilao, Batangas, farther south, they sport blue spots. Elsewhere, they're less bright and perfectly camouflaged in the reefs. Recent studies show that where chlorophyll a levels of the algae are particularly high, the mantle looks reddish brown; in contrast, if the mantle skin is rich in iridocyte levels compared with chlorophyll a, it shades to blue coloration. Because the colors vary by region, it seems likely to me that they can also vary with other components of a local site, such as light intensity or water clarity. The mystery that remains is understanding how and when and if the clam can selectively shift the abundance of chlorophyll a relative to iridocytes, and thus shift mantle colors under different conditions, ranging from blue to green and brown.

I continued to ponder the mystery of the different color spots on the clams as we swam across the giant clam preserve. I watched again, suspended above the bottom, mesmerized by the steady intake of water into the round, blue-rimmed incurrent siphon as one huge clam filtered water. This animal has been filtering water and photosynthesizing in the sun for a hundred years, as time spooled through the industrial age. It is not a sentient being, so can't be aware of its experience, but it must retain some stamp of the changing ocean and changing times in Bolinao. It is a strange type of time capsule, with the passage of the years and circumstance embedded in its vast shell. I thought more about the life cycle of this giant clam, so different from other clams. These gentle giants sit gaped open, a model for a sustainable life, filtering plant cells from the ocean water and using their solar power to transform light into food energy.

Once clams reach the giant stage, they should be home free and live on their reef for a hundred years, unperturbed by predators. But not with humans around. Humans learned how easy they were to harvest and how nutritious a meal they make. Although giant clams are huge, they are extremely endangered because they make such a valuable meal and are easily captured, but are long-lived and slow to recover. Giant clam is still a treasured food in many Pacific islands, prepared raw with lemon, simmered into coconut soup, baked into a savory pancake, or sliced and sautéed in a dozen other ways. Their adductor muscle is coveted as high-end sashimi and alleged as an aphrodisiac, and this, combined with luxury demand for their ivory-like shells, has driven *T. gigas* extinct in China, Taiwan, and other parts of their native habitat. Giant clams were virtually extinct in the Philippines in the 1970s. Because of their diminished population within the Coral Triangle, giant clams are listed as vulnerable in the Convention on International Trade in Endangered Species of Wild Fauna and Flora (CITES) and on the International Union for Conservation of Nature (IUCN) Red List. They have been brought back in some areas by reserves that also support ecotourism enterprises. That is why we were diving in a giant clam preserve, full of tiny outplanted baby clams and old giants that had successfully grown there. Some of the toughest marine-protection laws in the world, along with giant clam aquaculture, have helped Palau's and the Philippines' wild clams survive.

A newer threat can invade even the most well-protected reserves. Like reef-building corals, giant clams are susceptible to warming events that cause stress to their temperature-sensitive symbionts. When warmed, the symbionts are expelled from the clam, and like corals, the clam loses its color and bleaches and can subsequently die. Across the Pacific, as coral reefs are hit with warming events that cause widespread bleaching, giant clams are also succumbing to heat stress. The most at-risk reefs to lethal warming are the very shallow reefs like the one we were visiting.

The sanctuary near Bolinao is a joint project with the local government and Davao del Norte State College and is managed by the Adecor United Fisherfolk Organization (AdUFOr). The nursery is a model community project that balances livelihood needs alongside conservation, explained our colleague Ed Gomez, then director of the Bolinao Marine Laboratory. He pioneered the populating and restocking of giant clams in the Philippines in the 1970s when he imported giant clam larvae from the Solomon Islands and Palau, in an initiative supported by the Marine Science Institute of the University of the Philippines.

Giant clams have important ecological roles on a coral reef. They create habitat on reefs by building structures for animals and plants to live on and around. Restocking giant clams alongside coral transplants helps increase the abundance and richness of fish species. Giant clams are also efficient biofilters; they process huge volumes of water and improve water clarity and hygiene of degraded patch reefs.

THE GIANT CLAM'S LIGHT TRICK

The good news for giant clam conservation is that they are well suited to farming, because they produce such vast numbers of eggs that are rather easily reared to metamorphosis under the right conditions. The solar symbiotic algae, called zooxanthellae, are not embedded in the embryos, and so farmers must add them manually. At the Bolinao Marine Laboratory, they do this by isolating the zooxanthellae from an adult clam and then coculturing them in large batch cultures with day-four larvae. By day five, the larvae have ingested the algal cells and then transfer them via tubules extended from the gut to columns in their own skin. The algae continue to grow and divide, and populate the clam's skin for the rest of its life. The baby clams, superpowered by their solar cells, grow up to seventy-five times as fast by weight as some coral and much faster than other clams. They can grow up to two inches in their first year and reach football size and weigh five pounds by seven years.

After a long dive in the reserve, Morgan, Courtney, and I were tired and cold and ready to climb back in our boat. As Morgan handed up her scuba tank, she couldn't stop talking about how big and bright the clams were and wondered how old I thought the largest ones were. I knew that five-foot-long clams could be one hundred years old, so perhaps that four-foot one was sixty years old. We did know they had been growing there for at least forty years, because the reserve was initiated around 1980. As we soaked in the warm sun before motoring back, we talked about what a crazy underwater zoo we had just seen, with silent, nonmoving animals as long-lived as tigers and with urchins as keepers of a clean zoo in these overfished waters.

We also chatted about the ecological value of giant clams to a reef, especially here in an area of the Philippines where fish farming was creating murky water conditions. As long as they get enough light and the water is not too turbid, giant clams can clean up the water. Each giant clam is like a tiny water filtration system, sieving out excess plant cells and other organic matter and contributing to the clear water that coral reefs need. On the Great Barrier Reef, a sparse population of giant clams filtered over twenty-eight thousand liters of seawater per hectare, per hour. Listening to their excitement and observations, I thought to myself, "This is why I brought them." How pivotal this experience was for a PhD student, a future leader launching a new career in marine science, having had the chance to see a program like this internationally, being run at scale, with a cadre of Philippine students, professors, and volunteers who believed in the hope they were providing.

As we headed back to the lab to test samples of corals collected at different distances from the glut of fish farms, I pondered what we had seen. There is something poetic about the spreading light suffusing the skin of these gentle giants. They live for centuries, fixed in place on a sunlit reef like some huge redwood. Both giant clams and corals share this superpower of partnering with zooxanthellae to turn sunlight into food, but clams have taken the partnership to a new level by harnessing the algal cells into a solar reactor design. The quietly flashing iridescent colors that underlie a sophisticated solar reactor can transform and harvest light at wavelengths considered optimal by engineers. This reactor still holds mysteries in its design that are being investigated to optimize our own solar reactors. The essential new insights contributed by the clam's reactor

THE GIANT CLAM'S LIGHT TRICK

are ideas about how to design micro-reactors. Microarrays can be designed with columns of the tiniest solar cells, interspersed with reflectors to optimize generation of energy. In addition to contributing innovation in the design of tiny solar cells, the giant clam adds another solution for a world endangered by climate-induced flooding and pollution. This animal uses solar power to efficiently filter the water around it. In terms of inspired design and nature-based solutions, this one is a home run. Imagine fleets of giant clams or oysters bordering the shores of all our fish farms and tropical urban cities to grow ever-ready, constantly running, low-cost and possibly edible filters to clean the water. It really goes on and on full circle, because the solar-powered filtration capacity of giant clams can transform the reef around them to be cleaner and healthier and, with proper cultivation and innovative research, support the economy of rural communities in island nations like Palau, Indonesia, and the Philippines.

The multitude of eight-armed octopus forms include the common reef octopus at the top left, long-armed deepwater forms, the giant Pacific octopus, the ornate night octopus, and the long-armed, striped mimic octopus.

6.

The Octopus's Shape Shift

It started on a dive to a scroungy, poor-visibility back reef in the Philippines, with more dead than living coral. Our task in 2007 was to assess the health of coral in water that was cloudy with nutrients from nearby fish farms, and we had left the bright sunlit waters of the giant clam preserve for a nearby back reef. The day after our giant clam dive, Morgan, Courtney, and I dropped in from our boat and descended to the shallow remains of a once-thriving nearshore reef. A few small fish flitted above dead coral heads, which loomed like giant boulders from the murk. They had become substrate for sponges and turf-like growths of dark marine algae, both of which outnumbered live corals. The whole reef was grim and ghostly still, in sad contrast to the vibrancy, light, and color of the giant clam reef we had just visited. Unattached clumps of algae as large as soccer balls drifted by. Then movement a few feet away caught my eye. It was the same color and texture as the other algae growing on the dead coral, but the shape was wrong. It couldn't be part of the coral. Had I not seen the transformation happen, I would not have discerned that this particular clump of algae was in fact a small octopus, about the size of a football. I stopped swimming and hovered to watch it, happy to find something interesting on an otherwise depressing dive. It must have felt my gaze or heard my bubbles, because it swiveled, and our eyes locked briefly. Then it slowly lifted from its perch and, before my eyes, became a floating piece of algae, perfectly matched in color and texture and behavior to the other rafts of floating algae on this reef. This swift, smart, uncanny match was an otherworldly transformation that enticed me into the world of the octopus. And it worked—I rapidly lost track of this octopus as it drifted away with other algal clumps in the current.

The octopus in this sketch is showing a transformation from dark and bumpy on the dark reef to light-colored and smooth in the light.

This shift was quicker and more total than Clark Kent changing to Superman. The octopus altered the color of its skin from a uniform brown to mottled reddish brown; changed the texture of its skin from smooth to rumpled, with lots of ridges, bumps, and spikey projections; and adjusted its behavior from perching on the bottom to drifting suspended, just above the seabed with arms stretched into jagged, stiff, algae-like projections. This rapid shift is a superpower orchestrated directly by a dedicated part of the brain that activates a cascade of complicated muscular maneuvers. The nervous system control is called neural polyphenism and makes dramatic change happen fast. Scientists don't know how or when the initial plan of "look like algae" activates in the octopus brain. Somehow, in rapid succession, the octopus detected a potential predator, decided to shape-shift, accessed a stored blueprint for how to mimic an algal clump, activated innumerable, extraordinarily complex muscular contractions all over its body, and became something altogether different. The complete visible change can be accomplished in way under a second, in about two hundred to seven hundred milliseconds.

Octopuses are mollusks and thus relatives of giant clams and sea slugs. They are part of a larger group of quick-change camouflage artists, the cephalopods. Cephalopods also include squid, with the largest invertebrate in the ocean (the giant squid), the charming multihued cuttlefish, and the backward-swimming nautilus. The beginnings of the mollusks are hidden in deep time, but they are ancient invertebrates with an origin date of about 545 million years ago. The oldest cephalopod ancestors are a bit younger, estimated at 530 million years ago. These first cephalopods, the plectronocerids, swam like a squid but carried a tall, conical shell; they are believed to be the first swimming predators in the ocean, preceding predatory fish by about 200 million years. Dr. Sydney Brenner, the founding president of Japan's Okinawa Institute of Science and Technology, considered them the first intelligent beings on the planet.

Octopuses are distinguished from squid by having eight instead of ten arms, each lined with sticky suckers. The ten appendages of squid include eight arms that are similar to the eight suckered arms of the octopus and two extra long tentacles with suckers on the end. They also have three hearts to support an active, blood-pumping lifestyle as fast-moving underwater predators and a large brain and compound, vision-forming eyes to direct the activity. Technically they have nine brains: a larger central processor and an optic nerve in their body, and smaller clumps of ganglia in each of their eight arms. Octopuses are a very diverse group with over three hundred species and include the globally distributed common octopus (*Octopus vulgaris*), the giant Pacific octopus (*Enteroctopus dofleini*) in the Pacific Northwest, the highly venomous blue-ringed octopuses (*Hapalochlaena* spp.) from Australia, the caped blanket octopuses (*Tremoctopus* spp.) of midwater oceans, and the mimic octopus (*Thaumoctopus mimicus*) of the Indo-Pacific (all shown on the diagram on page 120). Despite diverse shapes, sizes, and habitats, all octopuses are quick-change artists. The core of the trick to change appearance instantly is shared by all octopuses: central brain control of skin color, skin pattern, and body shape.

All three components of change involve separate control by the brain and specialized neural control. A detailed understanding of the way in which the octopus brain controls body patterning still eludes us, but we know a little bit about the nuts and bolts of how the color and texture change works. While other invertebrates have color-changing cells called chromatophores, cephalopod chromatophores are different in that they are more complex and multicellular, essentially mini-organs under neural control. Skin color is changed by activating thousands of chromatophores, tiny pigment-filled sacs in the skin, each surrounded by a network of ten to thirty radial muscles controlled by neurons. When activated, radial muscles contract and expand the sac and reveal either red, yellow, or brown (in most species) pigments. Skin color changes in the shape set by the chromatophore muscles. When the muscles relax, the elastic sac compresses to its original size, returning the octopus to its starting color and pattern. Differently colored chromatophores are scattered in the skin and underlain by reflecting cells. The magic of the fast color change is conjured by the brain of the octopus. The activity of the chromatophores is controlled directly from a dedicated chromatophore lobe in the brain. In *Octopus vulgaris* there are over half a million neurons in the chromatophore brain lobes to activate (or inhibit) the different color classes of chromatophores.

The blue-ringed octopus (*Hapalochlaena* spp.) is a master performer in the Taylor Swift style; it not only changes the pattern of colors but also flashes shiny blue expanding rings. The extra-sparkly bling of bright blue in the rings of this octopus is caused by iridophores, cells with stacks of very thin crystals that reflect light in blue wavelengths—the same reflective cells used by the giant clam to optimize solar gain. When startled, the blue-ringed octopus changes to bright yellow with flashing iridescent blue rings. If threatened, it widens the blue rings in a threat display. Specialized cells called leucophores produce the solid background color. Gorgeous though this small octopus is, the flashing blue rings warn of a lethal toxin in a deadly bite, so I am just as happy we did not see it on our dive.

Of all the changes to the small octopus I saw on the reef, I was most dazzled by the instant change in skin texture from smooth to jagged and bumpy; this is also under a sequence of neural and muscular control. A multitude of pouches in the skin can erupt from smooth into jagged three-dimensional cones called papillae to produce bumpy skin. The collective action of radial muscles contracting the skin pouches pushes the cone of tissue outward, and internal muscles refine the shape. Each papilla has a white tip, which is produced by structural light reflectors (leucophores and iridophores) that lie between the papilla's muscular core and the skin layer that contains the pigmented chromatophores.

Both cuttlefish and octopuses perform this textural trick, but researchers have studied it most intensively in cuttlefish. Cuttlefish can attain a remarkable level of camouflage, and yet recent research shows part of the trick is having prepackaged color and texture patterns that can be used in different settings. Cuttlefish blend into a variety of backgrounds mostly using three main patterns. Uniform patterning matches the sandy ocean floor; mottled patterning matches larger pebbles or small drifting debris; and, finally, disruptive patterning matches large nonuniform objects in the background, such as larger rocks or debris. Quick change is needed, because a cuttlefish moves rapidly from one background to another. It can get complex fast, because cuttlefish also use color and pattern to woo potential mates and threaten competitors. A cuttlefish can even do both at once, in an essentially two-faced display. In each case, cuttlefish are able to coordinate color, pattern, and texture changes of their skin in as few as 125 milliseconds. This speedy skin change uses "visual assessment shortcuts," which are visual cues cuttlefish may see and use as an indication to activate a particular, prepackaged pattern style. While best studied in cuttlefish, these same processes control pattern and texture in octopuses.

Beyond skin color and texture, the deception is completed by the third component of a good octopus camouflage: the change in posture. Some species of octopus have particularly impressive behavioral displays that go beyond blending in with the background and blend in by standing out. Rather than masquerading as still, bottom-dwelling objects or organisms, they mimic the patterning and behavior of organisms in motion in the water column, and thus move around in their disguise. For example, the sand-dwelling Atlantic longarm octopus (*Macrotritopus defilippi*) mimics the swimming posture, style, and speed of the common sand-dwelling flounder. This behavior is especially useful for foraging in open sandy habitats, where there are fewer things to hide behind.

Perhaps the octopus species most acclaimed for mimicking other animals or plants is the aptly named mimic octopus. The mimic octopus is the master of quick disguise and makes sensational and unexpected underwater matches, just like that octopus from my dive in the Philippines. That octopus held its arms out stiffly, like jagged fringes of algae, and then mimicked the complex suspended drifting behavior of a mass of floating algae to wash away from me. Although our meeting was too fast for me to be sure what species worked this trick, the speed and perfection of the camouflage and the shallow, silty habitat suggest it was the famed mimic octopus. It not only matches the pattern of its habitat but actively mimics the movements and behavior of actual plants and animals. In addition to mimicking drifting algae, it can impersonate swimming flounder, lionfish, and sea snakes. In a video I watched, it starts out by turning into a huge, dark, and threatening form to scare a crab. Mission accomplished, it goes on its way, undulating in flattened form across the bottom, looking exactly like a swimming flatfish. This isn't quite good enough, so next it mimics a toxic, envenomated lionfish, complete with a multitude of spiny projections. Everything instantly changes in each new form, from its posture to its coloration to its swimming behavior. The final trick is to impersonate a skinny black-and-white banded sea snake. Strikingly, all three of the species mimicked by the octopus possess toxins: local flatfish have poison glands in their fins, lionfish have toxic-tipped spines, and banded sea snakes have envenomated fangs. Impersonating venomous or poisonous species is not an uncommon tactic in predator deception, but the ability to instantly shape-shift between different disguises is rare.

Octopuses have attracted fascination for millennia. Aristotle (384–322 BC) was one of the earliest fans we know about. Among his other accomplishments, Aristotle was a talented biologist and named over five hundred species of birds, mammals, fish, and invertebrates in his *History of Animals*, written in 350 BC. Most of his invertebrate observations came from his time on Lesbos in the Greek islands on underwater swims in the Pyrrha lagoon (now called the Gulf of Kalloni). It is mind-bending to read of his fascinations with octopuses and his observations of their denning behavior and how they change their color and form. He must have spent a lot of time watching them 2,500 years ago to observe what they eat, how they hunt, and how they age, as described in *History of Animals*:

> The octopus . . . will approach a man's hand if it be lowered in the water; but it is neat and thrifty in its habits: that is, it lays up stores in its nest, and, after eating up all that is eatable, it ejects the shells and sheaths of crabs and shell-fish, and the skeletons of little fishes. It seeks its prey by so changing its colour as to render it like the colour of the stones adjacent to it; it does so also when alarmed. By some the sepia [aka cuttlefish] is said to perform the same trick; that is, they say it can change its colour so as to make it resemble the colour of its habitat. The only fish that can do this is the angelfish, that is, it can change its colour like the octopus. . . . As a proof that they do not live into a second year there is the fact that, after the birth of the little octopuses in the late summer or beginning of autumn, it is seldom that a large-sized octopus is visible.

The biology behind the skin shifts of octopuses has baffled scientists for the thousands of years since Aristotle. It wasn't until the early 1800s that scientists investigated actual mechanisms behind this astounding ability. At first, scientists hypothesized that octopuses and other cephalopods changed their color through release of liquid pigment; however, it was soon observed that the various colors are housed in small spots in the skin, which change their size and form, but not location. In 1819, Giosuè Sangiovanni was the first to recognize these color-filled dots as specialized mini-organs embedded in the skin and responsible for cephalopod skin color change.

The next question was how chromatophores so rapidly produce different colors and patterns. Early researchers disagreed about the potential role of nerves and muscle fibers in the contraction and expansion of chromatophores. In 1882, French naturalist Raphaël Blanchard suggested that cephalopod chromatophores, while controlled by the nervous system, are made up of connective tissue rather than muscle fibers. He thought that cephalopod chromatophores had the same general structure as those in other animals including fish, frogs, and chameleons. However, other researchers of the time believed the fibers surrounding chromatophores were muscles and controlled by nerves. In 1892, Césaire Phisalix developed the theory of chromatophores being controlled by muscular fibers.

Finally, in 1901, Eugen Steinach proved Phisalix right with experiments showing that the chromatophore organs are controlled by the radial muscle fibers. Eventually, in 1932, Enrico Sereni and John Young showed that the chromatophores were also directly controlled by the brain. In the late 1960s, Ernst Florey and coworkers from the Department of Zoology at the University of Washington established a more complete working model of the chromatophores in squid as an organ with unusual brain connections and comprising five different cell types. Because Florey and Richard Cloney worked with chromatophores of squid at our own Friday Harbor Laboratories, I feel honored to be able to show their diagram of the structure of the cells in a squid chromatophore. The dark area in the center is the ink sac, surrounded by neurons to fire the radial muscles that enlarge the sac and make the skin appear darker.

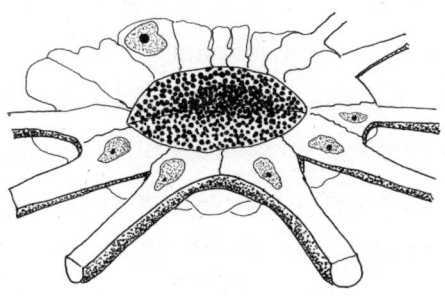

A squid chromatophore with inner pigment sac and radiating muscle fibers.

REDRAWN FROM E. FLOREY. *AM. ZOOLOGIST* (1969), 9:429–42

Investigation of the other cell types involved in color change, the leucophores and iridophores, didn't occur until the mid-twentieth century and was aided by improved microscopy. These also appear under neural control and can be adjusted and tuned for the desired reflective effect. Remember that iridophores are also the cells that sparkle in the skin of giant clams and change the wavelength of light that reaches the photosynthetic algae cells. In squid, iridophores can be tuned to change the wavelength of light they reflect and change their color. Similarly, some squid have adaptable leucophores that can change from transparent to reflective.

One reason I've delved deeply into the mechanisms of skin control in octopuses is that engineers and scientists are applying what we've learned about octopus skin to the design of high-tech smart materials. The smart skin of an octopus is a superpower that has dazzling capabilities. As far as humans are concerned, the ability of the octopus to change the color, pattern, and texture of its skin in an instant is otherworldly. We know a lot about how the shift is worked in the skin itself, but the way it is initiated by the brain and how the central brain interacts with the other eight brains in the arms remains unclear. Knowing how it's done doesn't actually demystify the wonder, because there is still more mystery than known science about how the magic is worked, especially from the perspective of a human who cannot do any of this.

A tricky part of reinventing the camouflage of an octopus for human application is its biological complexity. Both skin texture and color change are under electrical control by the nervous system, making them inspiration for smart materials. Smart materials are a big thing these days; they are those whose properties can be changed by external controls like temperature or humidity or electrical charges. Imagine a soldier's camouflage jacket that can change to match its background at the flip of a switch. Researchers from the University of Bristol have designed an artificial material that mimics just the color change by octopus chromatophores. An artificial skin made from stretchable material can effectively and instantly mimic the shade and color changes of chromatophores because it's under electrical control. At the flip of a switch that sends a charge of electricity, this material can change color and shade, and bands of color waves pass across the material, like the "passing cloud" display of an octopus.

Mimicking texture change in artificial materials is even more complex than color change. Engineers have studied how octopus skin can shift from smooth to bumpy in a split second. Recall that the papillae in octopus skin can change shape with muscle contractions under neural control. James Pikul, an engineer first at Cornell University and then at the University of Pennsylvania, studied the shape changes in the papillae of *Octopus rubescens*, a small California octopus. It has numerous skin papillae for camouflage and can extend a conical papilla 4 millimeters high in 220 milliseconds after activation, literally in the time it takes to blink an eye (200 milliseconds). He and his students produced a synthetic mimic of these texture changes.

Pikul and his team combined a stretchy silicone layer with an inflexible fiber mesh in predetermined shapes that could be inflated like a balloon to a 3D form. This work was initially funded by the US Army, which, not surprisingly, is interested in better ways to camouflage soldiers and tanks. The work is a leading venture in the science of smart materials and is now funded by prestigious basic research agencies. Their work with smart materials and surface texture also intergrades into programmable stiffness that can be controlled electrically. The key attribute of octopus smart skin is that it is under neural control, which is essentially electrical control, and triggers change in the blink of an eye. The skin of sea stars is also of interest in designing smart materials because of its strange properties under neural control, so I'll continue the discussion of engineering applications of programmable smart materials in chapter 8. These are both fantastic examples of the value of bioinspired design and the transformative advances in useful applications derived from the otherworldly innovations of invertebrates.

THE OCEAN'S MENAGERIE

All cephalopods, from octopuses to squid to cuttlefish, seem alien for their slippery shape-shifting, smart skin, and uncanny intelligence. Adding to the strangeness of their alien ways, many of the cephalopods are creatures of the night sea, and so it was on night dives that we went to find them. I was enticed into night dives in Hawaii and Indonesia between 2010 and 2016 with videographer David O. Brown to find and film living matches to Blaschka glass models of invertebrates. In the next chapter, I describe how the Blaschkas chose some of the most enchanting animals in the sea as subjects, from the ethereal jellyfish to bright-spotted nudibranchs to shape-shifting cephalopods. While the Blaschka glass collection includes multiple species of octopuses, squid, and cuttlefish, the array of octopus species is dazzling in showing more shapes, sizes, and capabilities than I have ever seen in life, from tiny to large.

The most memorable night dive with David was to a small coral reef near shore on the Big Island of Hawaii. When we first met, David's expertise was filming huge fantastic things like beluga and humpback whales with the Philippe Cousteau crew. But he has an eye for any kind of natural wonder and has long loved all kinds of cephalopods, from squid to cuttlefish to octopuses, so he had more skills than I in finding them on night dives. I didn't quite believe his calm certainty that if we did night dives in Hawaii we would see octopuses out patrolling. I was busy teaching a field course to Cornell students and was a bit grudging about freeing up my schedule from nighttime lecture writing to go on dives, but I was willing to

THE OCTOPUS'S SHAPE SHIFT

give it a try. We went with Denise Vidosh and Dave Rafalovich, owners of my favorite dive company, Blue Wilderness Dive Adventures. The four of us plus a boat captain headed out in the dark and soon reached a nearby site where Dave had seen octopuses. We dropped into the water with our lights and started poking slowly around the edge of the night reef. I was busy watching a bright pink flatworm crawl into a crevice when I heard the demanding, rapid tank clang. We can't yell to our partners underwater on a dive, so when we want to get everyone's attention, we bang on our tanks with dive knives, and the sound carries well underwater. This was an urgent, "come quickly" clang, either a cool find or a dangerous threat. I held my breath to listen and localize the sound and then swam fast go see. Dave and Denise had corralled a gorgeous spotted octopus (*Callistoctopus ornatus*), and David gestured for me to go close while he filmed it. I slowly moved in, amazed by the find. This was different from the wily, muscular day octopus I was to struggle with on a reef a few days later. This was a quiet, calm, serene creature. I moved in and easily picked it up. It perched calmly in my hand, and we eyed each other. Its eyes were large and luminous and met my own. It felt like being eye to eye with a human, except eerily different, because the shape of an octopus iris is more rectangular than round. I was completely lost in its gaze.

When I held that octopus and it sat calmly in my hands, a bond grew. I fell in love. I truly loved this animal and its epic quest to survive in our oceans. In that moment of love and empathy, I imagined I understood all it took to be an octopus: the challenge of creeping through the night reef to find prey like small crabs, the dangers of predators and the need to be on constant watch, the patience and obsession of caring for the eventual brood of eggs and tiny octopus hatchlings.

Now when I see an octopus or cuttlefish, I don't see just the animal but also their quest for a good life, and I want that for them in the same way most people want a good life for their children. They are real beings to me. It was a reminder that in addition to a global ecologist dealing with climate change and human impacts to once pristine oceans, I am also an invertebrate naturalist. These animals themselves are the source of my passion for sustaining our oceans. My bond with creatures in the ocean reverted to being heart-driven from being data-driven.

My twenty years as an underwater research scientist measuring the health of our ocean and its inhabitants had trained me to be an efficient data scribe. Ecologists can never have enough numbers and every dive seems too short, so we make the most of taking data on our slates or cameras on every dive. Seeing the miracle of the intelligent eyes that night gave me permission to watch and wonder underwater instead of obsessively writing numbers on a slate. These night dives to find and film octopuses marked a turning point in my career, so I think of the rapid skin shift of the octopus and its mysterious ways as a metaphor for my own big transition. It's nice that our hearts can sometimes override our brains as quickly as an octopus can change its skin.

Long before I became entranced by the wonder of cephalopods and nudibranchs, the glass artists Leopold and Rudolph Blaschka were obsessed with the rare beauty and strangeness of these animals. They elevated all the diversity of octopus relatives to masterpieces in glass art. Just consider that 170 years ago, they perfected the eye-to-eye experience with the most common octopus in glass. Then look at the heights they took their art in conveying the beauty and strangeness of all octopus relatives from squid to octopus to cuttlefish. As the curator of Cornell's Blaschka collection, I developed an online gallery of over five hundred glass images, and trolling through the images I can see ten different octopus species, fourteen different squid, and three different cuttlefish. I became inspired by the magic of the Blaschka glass artistry to go back and see even familiar ocean animals through the artists' eyes. This is also when I learned that art, like the Blaschkas' glass masterpieces, can transform our experience of nature. Our Blaschka models are ambassadors for nature in translating what some people see as a scary, slippery, slightly menacing animal into an object of beauty and wonder. Perhaps the best examples of elevating the frightening unknowns to exquisite masterpieces are both the octopus models and the jellyfish. Thus, we launched an offshore dive at night in the middle of the Pacific Ocean to find these animals.

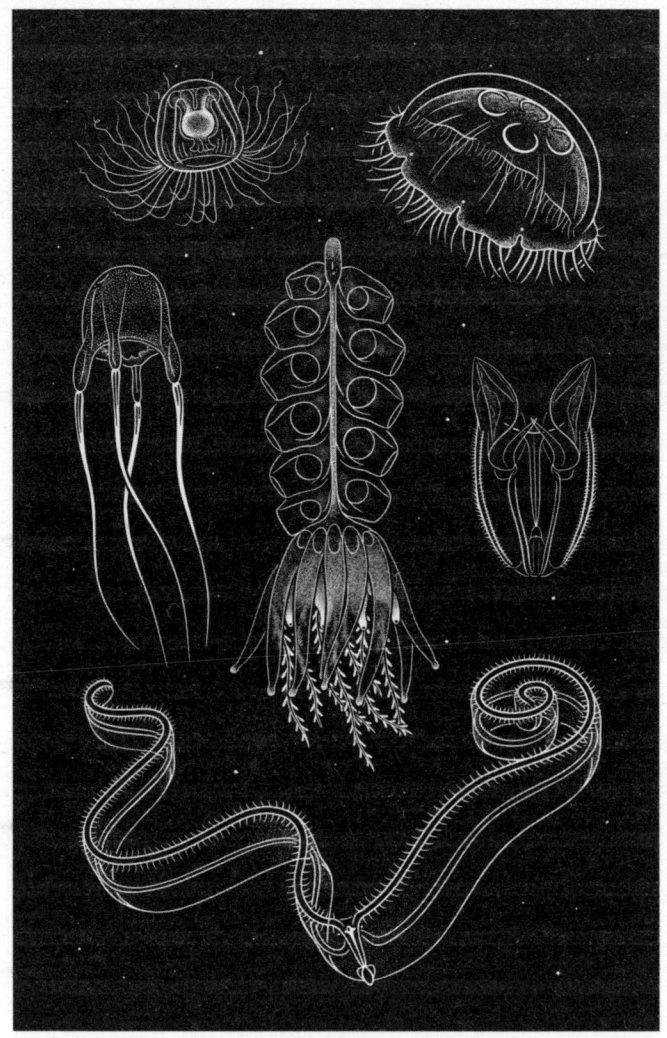

The myriad forms of jellyfish in the cnidarian phylum are depicted here, including siphonophores, high-belled hydrozoans, scyphozoan medusae, and cubozoan box jellies; and those in the ctenophore phylum, such as lobate and venus girdle forms. The siphonophore is central in this figure; moving clockwise from the top right are the scyphozoan, lobate ctenophore, venus girdle ctenophore, cubomedusan, and hydromedusan. Scientists usually call all of these gelata or simply jellies in recognition that they are not actually fish.

Despite toxic chemical defenses, marine sponges (like the red one here) are eaten by predators like nudibranchs. In a turnabout, sponges can attack living coral and take over their space.

A blue sponge overgrowing and killing a reef-building coral in Indonesia.

Admiring a yellow crinoid and an orange coral on an underwater cliff in Wakatobi, Indonesia, while searching for nudibranchs. PHOTOGRAPH © DAVID O. BROWN 2024.

An orange soft coral and a blue hydrocoral common on our coral health transects at Cennigan Wall, Indonesia.

Photographing a massive plating coral with tumors in Palmyra Atoll. PHOTOGRAPH BY BETTE WILLIS.

Flamingo tongue cowrie (*Cyphoma gibbosum*) feeding on the purple sea fan (*Gorgonia ventalina*) in Saint Croix.

Fingerprint cowrie (*C. signatum*) and flamingo tongue cowrie feeding on an octocoral with polyps withdrawn.

Closeup of the pinnate tentacles extended on polyps of the Caribbean octocoral *Plexaurella nutans*.

Hermissenda crassicornis from the San Juan Islands, an aggressive nudibranch that uses stinging cells it uptakes from cnidarian prey as its defense. PHOTOGRAPH © DAVID O. BROWN 2024.

A virtual circus of species, colors, and patterns common in these bright sea slugs. PHOTOGRAPH © NICHOLAS HESS.

Smooth-skinned octopus (*Benthoctopus lioderma*) photographed at Friday Harbor Labs.

A siphonophore jelly (*Rosacea symbiformis*) in glass, a likely match to the bioluminescent one seen on a blackwater dive in Hawaii. From Cornell University's Blaschka collection created circa 1880 as described in *A Sea of Glass*.
PHOTOGRAPH BY GARY HODGES FOR CORNELL.

Giant clam from Australia showing the characteristic gaping shell and brightly pigmented photosynthetic skin.

Crystal jelly (*Aequorea victoria*) photographed at Friday Harbor Laboratories where Osama Shimomura worked with the crystal jelly to discover chemicals in bioluminescence. PHOTOGRAPH BY CLAUDIA E. MILLS.

Hydrolab saturation diving laboratory with attached way station at a depth of approximately fifty feet in Salt River submarine canyon, Saint Croix, U.S. Virgin Islands. PHOTOGRAPH BY BRUCE NYDEN.

Aquarius saturation diving laboratory at a depth of approximately sixty feet in St Croix. The white, hexagonal structure called a way station contains a pocket of trapped air to allow divers to swim in and talk together. Aquarius replaced Hydrolab in 1986, which was decommissioned in the 1980s. PHOTOGRAPH BY JOHN OGDEN AND COURTESY OF HTTPS://DIGITALCOMMONS.USF.EDU/OGDEN7_IMAGES/93. © UNIVERSITY OF SOUTH FLORIDA.

The stars align in a tidepool in 2014 just before the massive sea star wasting epidemic killed millions of stars from Mexico to Alaska.

The author with the sunflower star (*Pycnopodia helianthoides*) as part of Friday Harbor Labs captive breeding program to restore a species endangered by the disease epidemic.
PHOTOGRAPH © DAVID O. BROWN 2024.

A year later in 2015, a sick ochre star (*Pisaster ochraceous*).

Cultured fragments of multiple species of coral newly cemented to the reef in a locally managed marine protected area in Fiji.

A nearby section of reef two years after fragments were cemented.

7.

The Jellyfish's Light Show

It felt lonely at day's end to leave the only bit of land in the middle of the Pacific Ocean. The next big landmass was 2,350 miles away in California; Japan was 3,000 miles away. The rocky shore and looming bulk of Mauna Kea volcano receded with the day's last light on a glassy ocean as we headed 2 miles off the coast of Hawaii's Big Island. Our dive site was an unnamed patch of open ocean, haunting in its emptiness in the growing dark. David Brown the videographer, my Cornell undergraduate student Catherine, and I suited up in the large rubber-pontoon dive boat, tossing on oceanic waves. It was 2016 and we were playing hooky from the Cornell ocean sciences field course I was teaching in Hawaii. When we first heard about black-water dives and saw the stunning photography of unusual animals drifting up from the deep and lit by their own luminescence in the dark, we felt the pull of adventure. We also thought that it could be downright scary.

Night dives anywhere can be spooky, but black-water dives in Hawaii are run offshore, in water thousands of feet deep. It's unsettling to think about diving unprotected above deep, unseen water in the dark, when big things can swim in quickly and where tiger sharks roam. Even scarier, according to reports, are nighttime attacks by cookie-cutter sharks that dart in and drill out a two-to-three-inch-diameter circle of flesh from divers or their normal prey. Those thoughts aside, we pushed on, hoping to see some of the rarer open ocean planktonic jellies and larval forms of octopuses and squid, which rise closer to the surface at night. As a marine ecologist, I have always studied sessile, bottom-dwelling organisms and their associates on reefs. What was I doing in the open ocean looking for free-swimming pelagic creatures? It was because I was motived as much by art as by science, and we were on the search for Blaschka matches, particularly among the jellies.

I thought about how taking this risk feels so counter to my basic instincts of self-preservation. The excitement of what we might see coupled with the strangeness of a new experience won out over my worries. Luckily, my two dive partners that evening were skilled and experienced. Catherine (now a PhD researcher/educator in Australia) was already an accomplished invertebrate biologist, not to mention a champion swimmer and a strong diver. So she was experienced and fearless enough to be a good extra hand on the dive, and it was a nice opportunity to broaden her skill as a research diver. This night, she was talking nonstop, torn between some nervousness about the dive challenge and excitement about what we might see.

THE JELLYFISH'S LIGHT SHOW

She was wondering how deep we would go and for how long. I explained that the animals we wanted to see migrate regularly all the way to surface waters at night because this is where the most food is; tiny phytoplankton photosynthesize all day, setting the table for deeper animals to surface at night when they can't be seen by bigger predators. So we would not go below fifty-five feet and thus had a full hour underwater. Catherine also remembered the beautiful lobate ctenophores we saw floating above the reef the day before and hoped we would see them here, because they are bioluminescent. This was such a nice thought, and yet we also knew that dives in the plankton are really a lottery, because each packet of water is so subject to winds and tides. So the ctenophores that were here yesterday could be on their way to Japan today. The mention of Japan reminded her of what we were about to do, and she asked what would happen if we separated from the boat in the dark. I didn't mention this was what had been on my mind. It certainly would not be good for any of us to be separated from the boat in the dark, because the wind could blow us far before anyone would start looking. We talked through our safety procedures and especially that all of us would be separately clipped into an underwater tether to the boat. And we would all stay very close together.

David is my long-term dive buddy, friend, and partner in our filming projects. He was smiling and quiet and in his happy place, not at all worried, about to be underwater and also hoping for some strange new bioluminescent jellyfish, and any kind of squid.

After double-checking all our gear, we dropped over the side of the Zodiac into cold, black water with no sun to warm us. With upward of two thousand five hundred feet, or a half mile, of water below us, there was no way to anchor, so we clipped into safety tethers on the free-floating boat. Those sixty-foot tethers were our lifelines, because if we separated from the boat in the currents, the increasing wind, and the dark, we would quickly be lost at sea. We each had flashlights and cameras, and after an OK sign to one another, we slipped deeper down along our weighted tethers like secret agents from *Mission Impossible* scaling a wall. This was my first black-water dive, and as we descended, I felt strangely disoriented. In the total darkness and with no pull of gravity, I had no way to know what was up toward the surface or what was down toward really deep water. In the dark, I could not even see my bubbles rising for orientation. All I had was my hand on the tether.

We descended past twenty, thirty, and forty feet, watching our gauges carefully as the only reference for depth. The air in our buoyancy vests compressed with depth, the water got colder, and we descended ever faster. At the same time, we could feel the current's pull as the surface wind pushed the boat and took us with it. I strained to look outward and below, but it was dark in all directions, and my feeling of vulnerability and disorientation was almost overwhelming. I nervously checked my depth gauge, mindful to stop at fifty feet and hoping my student also had her head together. I needn't have worried, because she was a fish in the water and having the time of her life, sending over a happy thumbs-up to David and me. Then we were there, dangling like lures on a fishing line, lights of my dive buddies a comfort in the huge dark of the night sea. Our lights would obscure any bioluminescence, so, as if on signal, we three doused them and floated quietly in the dark. Waiting. Small specks of light

THE JELLYFISH'S LIGHT SHOW

like distant stars started to wash toward us. These were tiny phytoplankton and zooplankton, no bigger than a ladybug, but still lit by their own luminescence. We were immersed in a continuous stream of tiny blinking lights. Then, at the edge of my vision, a large glow floated toward and past me, too far to identify.

Another drifted slowly closer, and my light revealed a four-inch-long lobate ctenophore. This animal is a swimming light show, with both bioluminescent blue light glowing from cells lining the edge of the lobes and sparkling iridescence from its rows of fluttering ciliary hairs. The glowing light in this ctenophore comes from cells that underlie eight flashing rows of cilia. Despite such exquisite beauty, this is a fierce oceanic predator, riding the currents of the open ocean and actively hunting small crab relatives (shrimp, crab larvae, copepods, and ostracods), like some predatory space alien. This gorgeous ctenophore was likely *Leucothea multicornis* and may well have been the same species as the one Catherine saw the day before above the reef near shore. Then a larger glow washed from the dark and, as it came closer, spread into many smaller, attached lights. It was a chain of colonial salps, looking like fifty glowing, barrel-shaped, swimming stomachs gliding past us. A salp is a modified sea squirt or tunicate, and although it is called a jelly because it's a transparent plankton animal, it is more closely related to humans than to cnidarians. The current was coming faster now, because we were being pulled more quickly through the water as the surface wind on the boat had picked up. Then I saw an odd bluish glow and pulled at the edge of my tether to get closer to intercept this new form. There it was! A match to one of the glass models in our Blaschka collection.

THE OCEAN'S MENAGERIE

As a young assistant professor just arriving at Cornell to teach students and run research, I was also handed the job of curating Cornell's Blaschka collection, which includes exquisitely rendered cnidarian jellies, ctenophores, nudibranchs, squid, and octopuses. Our historic collection, ordered on approval in 1885 by Andrew Dixon White, Cornell's first president, includes over 573 invertebrate glass models created by the father-son Blaschka team. The day I first saw these invertebrates in glass, perfectly capturing the form and gestalt of many different kinds of soft-bodied invertebrates, I was hooked for a lifetime of preserving and studying the collection. While over sixty universities worldwide have Blaschka invertebrate collections, Cornell's is the largest. Leopold and Rudolph Blaschka are the same glassmakers who later crafted Harvard's glass flower collection. Saying yes to curating Cornell's Blaschka glass invertebrate collection propelled me into a new world of art, museums, filmmaking, and books. Our collection is a time capsule, showing invertebrates that were common in the oceans of the late nineteenth century. Leopold Blaschka created the collection to depict wonders of invertebrate diversity to his world in 1870. He could never have imagined that the fragile glass would endure when the living biota is being threatened in our time of changed climate and ocean. I know he would be proud and stunned to see his legacy still relevant in our more modern world. The Blaschka collection inspired my quest to find and show the living matches in our modern ocean as described in my first book, *A Sea of Glass*, and David Brown's and my film, *Fragile Legacy*.

THE JELLYFISH'S LIGHT SHOW

To explore how many of the creatures we could find, I dove around the world at locations rich in Blaschka matches: Mediterranean Italy, the Pacific Northwest, and islands off the coast of Maine, Hawaii, and Indonesia. I also worked with my students through checklists of living invertebrates. This quest is chronicled with brilliant photos of the glass and live forms in *A Sea of Glass*, but the very short summary is that some of the matches are endangered in today's ocean, particularly in the Mediterranean. Some of them are very rare, but many of the common forms still exist, and we could not actually confirm any extinctions in the group. So I think of this as a message that our oceans are still full of an amazing bounty of richness, even though perilously endangered by a dangerous mix of overfishing, climate change, and pollution. The most spectacular of the Blaschka models are the ethereal and delicate jellies, so that night we were on the hunt for rare deepwater jellies that migrate to the surface at night. The only way to see these delicate beauties is to jump in the ocean and look, because they are too fragile to be collected by nets and plankton tows or maintained in most aquariums.

The Blaschka jellies are magnificent in their craftsmanship and biological detail and exist in an astonishing variety. Perhaps in analogy with the animals' ancient evolutionary origins, the Blaschkas started by crafting single-belled jellies in so many shapes, sizes, and colors, some with only a couple of tentacles, and others with cascades of them. Even the simple, single-belled jellies are exquisite masterpieces, requiring the crafting of extremely thin bells in glass, as well as delicate tentacles, adorned with spots, speckles, and colored eyespots.

THE OCEAN'S MENAGERIE

The Blaschkas' glass siphonophore,
Apolemia uvaria, *from the Cornell collection.*

KENT LOEFFLER, PHOTOGRAPH COURTESY OF
CORNELL UNIVERSITY BLASCHKA COLLECTION

The cnidarian jellies are distributed into three different classes that superficially look similar but have different evolutionary ages, sensory systems, armatures, and swimming abilities; they are more different from one another than a spider is from a butterfly. They are distantly related to corals but diverged to their own evolutionary path over 500 million years ago. As open ocean voyagers instead of bottom dwellers, jellies need entirely different sensory systems and muscle types for locomotion. Genetic differences among some of the groups of jellies are as great as the difference between a human and a sea urchin.

THE JELLYFISH'S LIGHT SHOW

The oldest group containing jellies, the hydrozoans, are most closely related to corals and retain a coral-like, bottom-living form. These hydrozoan jellies tend to be small and fast moving with tall bells. The exception is one spectacular subgroup of the hydrozoans called the siphonophores, which I think of as the pinnacle of jellyfish complexity.

The siphonophores include the notorious Portuguese man-of-war, with potent neurotoxins that are strong enough to kill a human. Evolution has gone free-form in the design of siphonophores, by innovating multibell forms with many variably shaped and sized swimming bells and bright, venom-filled dangling tentacles and innovative sensory capabilities. The Blaschka collection includes some of the most complex and spectacular of these forms, such as the Portuguese man-of-war and the hula skirt jelly, each life-size and approximately eight inches tall.

While the siphonophores are the most complex, the best-known cnidarian jellies are the mighty oceangoing scyphozoa, some larger than a human and spending most of their life at sea. But even this group retains its evolutionary ties to the bottom of the sea, in the form of a life stage called a polyp, which buds off the oceangoing jelly. The most dangerous group of these jellies are the boxy-looking Cubozoa, an evolutionarily more recent class related to the group of the scyphozoa, with potent, deadly neurotoxins. That night was my first look at tropical jellies, and I hoped to see complex siphonophores and large scyphozoans but not see any swarms of dangerous cubozoans. I thought we might also see diverse forms of ctenophores, which superficially look like they belong to the cnidarians but hide a fantastic secret in their DNA and neurons.

Jellyfish, gelata, and *jelly* are general terms that include transparent, plankton-dwelling animals from diverse nodes on the tree of life. Many of the jellies I'll talk about are coral relatives in the larger group Cnidaria, which all produce stinging nematocysts. But the ctenophores are a look-alike group that the Blaschkas captured and that we see often. They were once considered close relatives of the cnidarians, but they do not have nematocysts, and new research posits that they may in fact be the sister group to all other animals and only distantly related to cnidarians. As one of the most ancient of all invertebrates, estimates put the origins of their ancestor at over 600 million years ago. New work confirms the surprising conclusion that they may be the deepest branch on the tree of life, and sponges, the other contender for deepest branch, are a newer offshoot. The new work concludes that ctenophores are different from all other animals for two reasons. First, large blocks of DNA sequences are shared between sponges and all other invertebrates except ctenophores. Second, ctenophores have a distinct nervous system; it is a single conducting web with no synaptic junctions, like other animal nervous systems. The ctenophores come in three basic shapes: the lobate, the tentaculate, and the venus girdle. Instead of venom-filled nematocysts like cnidarians, ctenophores have sticky cells with entangling threads for prey capture. Almost all the cnidarian and ctenophore jellies are carnivores, hunters that ply the open ocean for tiny planktonic prey that they capture in their entangling or nematocyst-filled tentacles. Some cnidarian relatives of the corals are armed with stinging nematocysts so powerful they can even take down a human.

THE JELLYFISH'S LIGHT SHOW

It floated steadily toward my light and I thought, "This is why I came." It was another ctenophore, but a rare, deepwater form. It was my first ever sight of a six-inch-long linear ctenophore, called Venus' girdle (*Cestum veneris*), swimming like a giant, transparent, flattened worm through the water. But this worm was lit up with a blue glow along the edges, topped by iridescent flashes of the glittering comb rows as it undulated so slowly through the water. This alien being is a voracious predator, swimming through the ocean, hunting for small crustaceans and pelagic mollusks. It's a match to the three-inch glass figure in the Blaschka collection. This was a small one, but they can reach six feet across. So just like an avid birdwatcher adding a new line to a life list of bird species, I added a new line to my life list of invertebrates, sharing kinship with my Blaschka teachers. The glow from a ctenophore is especially notable, because this descendant from the most ancient of animals actually has the genes to make its own bioluminescence, as I'll describe later, while most cnidarian jellies have to steal the light-producing chemicals from their prey.

Caught up in the excitement of discovery, and relaxed in my ocean redoubt, I completely forgot we must have looked to a shark like lighted bobbing lures on a fishing line or that there was even a boat somewhere above us. I was an ocean dweller, in my happy place of looking for the biodiverse wonders of the night plankton. Time disappeared and the known world receded as we dwelled in the deep. Another glow appeared and reminded me of Leopold

THE OCEAN'S MENAGERIE

Blaschka's words, reprinted in *A Sea of Glass*, as he watched the night bioluminescence a hundred and seventy years ago from his ship's deck while crossing the Atlantic.

> We are on a sailing ship in the Atlantic Ocean, immobilised because of the calm; it is a beautiful night in May. Hopefully we look over the darkness of the sea which is as smooth as a mirror: in various places there emerges all around a flash-like bundle of light beams, like thousands of sparks, that form true bundles of fire and of other bright lighting spots, as if they are surrounded by mirrored stars. There emerges close before us a small spot in a sharp-greenish light, which becomes ever larger and larger and finally forms a bright shining sun-like figure. A second one develops, a third; ten, a hundred of these suns light up at a certain distance from the peculiarly sparkling intervals, bright lighting circles form strangely formed figures, with in between places in a glowing light, an indescribably beautiful scene originates. . . . It is, as if they wanted to lure the enchanted observer into a realm of fairies.

Another small spot of light in our realm of fairies turned into an exquisitely shaped siphonophore (a cnidarian) in the light of my flash, its fishing lures spread out with blinking lights. This is a smaller cousin of the great Portuguese man-of-war and a match to one of our most sensational Blaschka glass jellies. This one was small with its main swimming bell about an inch high, but its lures spread well over a foot in all directions, casting an explosive net to ensnare small zooplankton. This siphonophore jelly is a remarkable life-form that has been called a superorganism, because it is evolutionarily developed from a single jelly to a colony with sometimes hundreds of units, with those units separated into groups with different capabilities. Some feed, others reproduce, and still others are armed for defense. While they look simply like a bunch of medusa bells tied together, they have a nervous system that coordinates the units. If we take a closer look, it starts simply enough at the top with a rather standard swimming bell, and sometimes that is all you see. But attached is a long tentacle trailing modified swimming bells that no longer swim but instead are heavily weaponized. The toxins packed into the harpoon-like, explosive nematocysts of some siphonophores are potent enough to kill a person and easily subdue even fairly large fish. Some siphonophores are over one hundred feet long with a trailing curtain of explosive tentacles. Indeed, probably the greatest danger to us on this dive would not be a shark but rather a large siphonophore like a Portuguese man-of-war. If an aggregation of the man-of-war came through, we would need to immediately surface to the boat, because the myriad tentacles with a deadly sting would be invisible in the dark. This was on my mind, because we had a bad encounter on a dive in Mexico, described in *A Sea of Glass*.

This one flowed by too quickly for me to be certain of the species, but it looked like our Blaschka *Rosacea cymbiformis* and was likely the species common in Hawaii, *Praya dubia*. It is a deep-sea dweller, rarely seen during the day, but it migrates vertically thousands of feet to food-rich surface waters at night. Large ones can reach up to one hundred thirty feet long. It is an active swimmer and attracts its small zooplankton prey with bioluminescent lights. The blue glow from the main bell lights its path through the water. The fishing lures would be invisible during the day but are lit in a curtain of tiny blue lights in the dark. This is a wonder that is rarely seen, because it lurks in the deep ocean during the day and only rises up at night. I've seen it reported previously only from submersible dives from the Monterey Bay Aquarium Research Institute.

As it came near, it was bumped by something in the water, gave a bright flash, and then disappeared. The foot-long chain of fishing lures was zipped up and the swimming bell activated for escape swimming. The reason siphonophores are sometimes called a superorganism is that the many units of the colony are variations on the original theme of the swimming bell, so in that sense, they are all like separate animals tied together in a chain. But they behave like a proper, integrated animal because they are linked together with an impressive and highly functioning giant neuron and muscle that instantly retracts the chain of polyps and activates escape swimming by the main bell. The evolution of this kind of hierarchical organization in a seemingly primitive invertebrate has fascinated biologists since the time of Ernst Haeckel.

Haeckel was an intellectual giant of his time who lived from 1834 to 1919 and was a contemporary of Darwin. He was particularly fascinated by jellyfish, which he called "crystal animals" or "glass animals." He was also a contemporary of the Blaschkas, who were inspired by his art and science and corresponded periodically with him. Fascinated with siphonophores, Haeckel advanced his theories of the connection between development and evolution, thus anticipating the now well-developed field of evo-devo, which infuses the study of evolution with developmental biology. His monograph on the siphonophores shows how the wonder of a siphonophore like *Physophora* (the hula skirt jelly) develops from its larval stage. His artistry and scientific insight were copied in glass by the Blaschkas, who not only crafted a magnificent adult *Physophora* but also reproduced for students the larval stages in glass. These are all in Cornell's collection.

Time passed unwatched as we hung silently, breathing on our tethers, bubbles rising invisibly above us. We watched the jewels of night plankton sparkle by. We were attentive hunters, eyes straining to see the next mystery glow emerge and take shape in our lights. It suddenly occurred to me to check my air; I was startled to notice my tank was nearly empty, almost an hour had passed, and I was shaking with the cold. It was time for us to leave this underwater redoubt and ascend to our surface world. I flashed my light three times and pointed a thumbs-up, "let's ascend" signal in my light. We ascended as a group to fifteen feet and stopped to hang as a threesome for three minutes on our safety stop, a precaution to allow any extra nitrogen to bubble out of our bodies. It was a quiet time as we floated motionless in the dark, processing a rather mystical experience we had just shared, silently reconnecting by eye contact.

After three minutes, we surfaced beside the boat, unclipped from our safety lines, and handed up dive weights and tanks, talking all at once about what we had seen. We'd spent an hour underwater, seeing amazing things, and because we didn't dive with radios, this was our first chance to talk about it. Catherine had the catch of the night. She had seen a tiny baby octopus, no bigger than a caterpillar, surfing by on the current. David and Catherine had also seen the Venus' girdle and the salps and lobate ctenophores, but had missed the siphonophore's light show. The chance to see that delicate siphonophore in its natural habitat, a remote patch of open ocean, was phenomenal. It did my soul good to see it calmly traversing ocean waters for its prey, flashing bioluminescence. This living Blaschka match connected me with Leopold Blaschka; I was briefly a nineteenth-century naturalist on a strange and romantic adventure to search out biodiversity.

The alternation between night and day has been a constant since the origin of life, and despite being something we take for granted, it is a fundamental background condition of existence for our spineless menagerie. It is dark about half the time in the shallow ocean, or all the time in the abyssal zone. So it is not so surprising that the ability to bioluminesce is the norm underwater and has evolved many times in the invertebrates.

It is estimated that over 70 percent of underwater invertebrates can bioluminesce. This evolutionary ubiquity signals to scientists that underwater light is both functionally important and easy to evolve. The problem is that the evolutionary origins of light are hidden in the shadows of early evolution. Dr. Steve Haddock of the Monterey Bay Aquarium Research Institute suggests that luminescence did not originate until well after the Cambrian explosion of life and first appeared in the crablike prey of jellyfish, perhaps 400 million years ago. This and other estimates puts the origin of light generation between 400 million and 500 million years ago. For perspective, insects on land like fireflies might have evolved bioluminescence much more recently, approximately 100 million to 140 million years ago.

Of all groups in the sea, the cnidarian and ctenophore jellies stand out for their mastery and diversity of light production; between 90 and 100 percent of these groups bioluminesce. Bioluminescence is a key superpower in the evolution of cnidarian jellyfish, and the siphonophores take it to an extreme level. The siphonophores are all in as a bioluminescent group; on an evolutionary tree, every single group of siphonophores bioluminesces. Despite the ubiquity of luminescence in the jellies, it is interesting that cnidarians do not have the genes to make the key bioluminescent proteins, and thus the ubiquitous bioluminescence in these siphonophores must have originated in crab-relative prey and was later co-opted by the jellies.

Bioluminescence is relatively rare on land, because only few invertebrates, like fireflies and glowworms, emit light. This contrasts with the high numbers of bioluminescent invertebrates in coastal and deep ocean waters. The number is even higher in the deep sea, where most of the animal kingdom emits, flashes, or glows light: squid, octopuses, shrimp, single-celled organisms, fish, sharks, rays, and jellies of all kinds. Dr. Kevin Raskoff of California State University, Monterey Bay, estimates that 90 percent of all the animals in the deep sea bioluminesce in one way or another. This light has been appreciated also on land for over two thousand years, when the Roman scholar Pliny the Elder noted that he rubbed the slime of *Pulmo marinus*, a jellyfish from the Bay of Naples, on his walking stick to "light the way like a torch." The ubiquity of bioluminescence raises the question, What is the function of all this glowing underwater light?

Bioluminescence is a predominant mode of communication in the ocean, and there is a lot to flash about. Topics to communicate by bioluminescence mostly fall into the category of repelling predators or attracting mates, but there are also cases like the lantern fish and the viperfish, where light is used to lure prey. Some animals use it to attract mates. A male seed shrimp (an ostracod, *Vargula hilgendorfii*) will squirt out a bright dot of light, zip upward, and then squirt another and another, essentially drawing an arrow that points out his whereabouts.

As a graduate student living underwater in NOAA's saturation diving laboratory Hydrolab (described in chapter 4), I had nightly watched the seed shrimp mating spectacle. Once the sun set and full darkness descended, my favorite part of watching life underwater—the dance of the seed shrimps, each as tiny as a sesame seed—began. First a few small flashes of light sparked. And then another and another. A column of tiny lights seemed to grow upward from the reef and sway in the gentle deep waves near the bottom. Light spots rolled across the reef as countless tiny seed shrimp males, each one encased in a clam-like shell the size of a sesame seed, contributed their display in hopes of attracting a female. The shrimp were as punctual as a train schedule. We could be in position every evening at the exact time to watch their mating dance unfold. Researchers like Jim Morin went on to do the pivotal work to show how tropical seed shrimp use their light in a mating display that rivals the enchantment of fireflies on land and varies spectacularly among species. The seed shrimp's light show is similar to that of fireflies in the small size of individual lights and the ability of the seed shrimp to make its own luminescent proteins,

THE JELLYFISH'S LIGHT SHOW

but differs entirely in its fluidity underwater. The male emits a small pulse of a viscous, bioluminescent fluid that glows in the dark and floats in place for up to ten minutes in the water. A single male can swim upward and create a briefly enduring pattern of vertical light dots. A population of males can stage an entire light show, and different species produce diagonal or horizontal lines of lights instead of vertical. Either way, it's a glowing road map for a female to follow to find her mate.

Instead of attracting mates like the seed shrimp, other creatures use bioluminescence to lure prey. The viperfish (*Chauliodus sloani*) dangles a luminescent lure in front of its mouth and then gobbles any creature that investigates.

The animals that flash light to fend off, surprise, or trick predators are the most common. The deep-sea shrimp (*Acanthephyra purpurea*) vomits bioluminescent goop into the face of threatening predators and then jets off in escape. Other animals flash light as a defensive burglar alarm. They light up to attract a second predator that will eat the first one (or to make the first predator think that a second one is coming, and so prompt it to leave). Although bioluminescence has exploded across the underwater tree of life into uses from mating to attracting prey to scaring predators, the chemistry of making light is similar.

In nearly all light-emitting organisms, from jellyfish to sea stars, bioluminescence requires three ingredients: oxygen, a light-emitting pigment called luciferin (from the Latin word *lucifer*, meaning "light-bringing"), and an enzyme called luciferase. When a luciferin reacts with oxygen—a process facilitated by luciferase—it forms an excited, unstable compound that emits light as it returns to its lowest energy state. This activated compound is triggered to flash by small quantities of calcium.

In cnidarian jellies like the siphonophores I saw in Hawaii, either an enzyme, cnidarian luciferase, or a photoprotein (for example, Green Fluorescent Protein) is added to a luciferin substrate called coelenterazine. These two chemicals must meet and react with oxygen to produce light, but the light is not generated until it is catalyzed by yet another element entering the light-containing cell. The control by a cofactor as a trigger provides a way to precisely time the flash of light and control the exact amount, kind of like a dimmer switch. The presence of a cofactor made the science to unravel the process very tricky to uncover.

The chemistry of light generation across all animals gets very complex, but the story within the cnidarian jellies is the most extraordinary in terms of the widespread use of bioluminescence in dark ocean waters and the story of the remarkable evolution of light and diverse light-producing proteins in jellyfish. All the jellies use coelenterazine as the substrate (the luciferin). Within the jellies, one group (Hydrozoa, including the siphonophores) uses photoproteins, and another (Scyphozoa) uses enzymes called luciferases. The fact that there are these two different avenues to activate light production is meaningful to biologists because it suggests these two pathways are independently evolved; it's further very meaningful to humans, because both these light-producing chemicals are now vital as fluorescent biomarkers in all biomedical research. A brilliant scientist discovered both of these and won a Nobel Prize for the discovery.

THE JELLYFISH'S LIGHT SHOW

As a graduate student, I used to watch a tall, slim Japanese man, Osamu Shimomura, pay kids at the Friday Harbor Labs a penny a jelly to collect hundreds of the crystal jelly (*Aequorea victoria*) off the dock. He would then patiently isolate the bioluminescent chemicals from these jellies until he had a large enough volume of material for chemical analysis. It took years. Dr. Shimomura was obsessed for his entire career with the chemistry of bioluminescence from his beginnings in Japan working with the tiny seed shrimp (*Vargula hilgendorfii*) to his now famous work with jellyfish. From 1961 to 1988, he and his colleagues collected over a million jellies to extract the light-producing chemicals in their research on the chemistry of bioluminescence.

Shimomura survived the blinding flash and radiation of the atomic bomb in Nagasaki as a high school student. After the war, he began a lifelong quest to uncover the chemistry of bioluminescence. He began his work in Japan with tiny seed shrimp. Next, on a Fulbright fellowship to Princeton, he teamed up with professor Frank Johnson to work at Friday Harbor Labs in the 1960s on the bioluminescence of the crystal jelly. To discover and isolate the light-emitting chemical, Shimomura estimates that he collected over a million crystal jelly specimens. His early work was slow and frustrating because none of his isolates separated from the jellyfish would luminesce, but a fortuitous small accident in the summer of 1961 showed the way forward.

He had discarded one of his nonluminescent isolates into a lab sink, puddled with seawater. He was shocked with an unexpected bright flash of blue light. Further tests revealed that he had in fact isolated the right photochemical all along, but no light emerged because he lacked the right pH and the calcium from seawater that catalyzed the reaction. It took another year of long days and nights in the laboratory to fully isolate and describe the chemical he eventually called aequorin. He discovered a complex chain of events that underlies the bioluminesce of the crystal jelly, which has special cells containing the photoprotein aequorin and its enzyme. To produce a flash of blue light, calcium needs to be added to the cell.

Over the next decade, Shimomura and colleagues showed that aequorin was a large chemical but contained within it a smaller chemical called coelenterazine. Here's the kicker. He had seen coelenterazine long before! Finding coelenterazine as a component of aequorin stunned him, because coelenterazine was the same bioluminescent chemical he had discovered so long ago in the seed shrimp in Japan. This lucky convergence of finding the same key chemical in both animals would eventually also connect the dots to solve the mystery of who makes the bioluminescent chemicals. It turns out that jellies do not have the genes to synthesize coelenterazine. They need to eat seed shrimp or other crustaceans to acquire coelenterazine in constructing the larger aequorin molecule. Over evolutionary time, jellies have co-opted the chemical pathway of their prey but never evolved the actual genes to make it. And like a conjurer with multiple tricks, the story does not end here, because the crystal jelly does have another startling capability.

Crystal jellies make more than one kind of light. They flash blue light when bumped, via the binding of calcium to aequorin. But they also glow green via a different pathway. Shimomura was laser focused on the isolation and purification of aequorin in 1961 and certainly had his hands full of big discoveries, so he didn't follow up on the isolation of the green photoprotein until much later. He simply called it green protein in his early research paper because its purified form was green under room light.

My colleague from Cornell, Dr. Jim Morin, was a graduate student working on bioluminescence when Shimomura was working in Friday Harbor. Jim was the one to name this green glowing protein green fluorescent protein. Shimomura continued the purification and crystallization of GFP and next described its structure. GFP itself is a large molecule with over twenty-eight carbon atoms in its skeleton. The chromophore is the part of the chemical that gives off the green light and is composed of ten carbon atoms, attached to the larger molecule. In 2005, Shimomura published a report on the new GFP photoprotein. Two other scientists, Martin Chalfie and Roger Tsien, published amino acid sequences, cloned the genes, and made GFP available as a tool to track anything you want, since the cloned genes could be incorporated into other organisms. Together in 2008, these three scientists were awarded the Nobel Prize in Chemistry.

While all jellies and their coral relatives use coelenterazine as the substrate to generate light, it is baffling that scientists have not yet found evidence that they have the genes to make it. Instead, the coelenterazine is acquired by eating crustacean prey and then carbon atoms are added by the jellies to create the larger molecule aequorin. Given how widespread bioluminescence is in jellies and other cnidarians, it is earth-shattering to me that they cannot synthesize it and must be constantly acquiring it from their prey. Talk about the "ties that bind" in this vital link and transfer of a required chemical in the food chain.

So why are the discoveries of aequorin and GFP so valuable for humans, and why does the discovery of GFP rate the highest scientific honor on our planet, a Nobel prize? These are separate and hugely valuable fluorescent chemicals for biomedical research. Aequorin is on a hair-trigger to emit bright light and the trigger is pulled by even the tiniest amount of calcium. Because the brightness of the light depends on the exact amount of calcium, aequorin has developed into an essential biomedical tool to measure the amount of calcium in a cell. The level of calcium is key because its increase triggers all sorts of vital human cellular processes, including contractions of cardiac and skeletal muscles, blood clotting, bone mineralization, and the smooth functioning of the nervous system. The measurable brightness of aequorin is a tool to study the role of calcium signaling in the progression of cancers and neurodegenerative disorders. Calcium signaling studies are also used to test responses to drugs prior to clinical trials.

It is stunning that Shimomura's work with the crystal jelly led to a second, even more valuable fluorescent label, Green Fluorescent Protein (GFP). It stars in hundreds of top scientific and medical papers because it reveals previously unknown actions of specific genes. As Roger Tsien put it in his work with GFP, "Our work is often described as building and training molecular spies, molecules that will enter a cell or organism and report back to us what the conditions are, what's going on with the biochemistry, while the

THE JELLYFISH'S LIGHT SHOW

cell is still alive." The tasks these "molecular spies" take on are to mark cells with fluorescence, so they can be tracked.

GFP marks and tracks a remarkable array of cell types and processes, from developing brain cells, to cultures of pathogens, to nematocyst pathways in nudibranchs. It is considered the most powerful tool to monitor gene expression in different kinds of cells. For example, in transgenic organisms, the transformed genes can be tagged with GFP and tracked visually during development. GFP can also tag cancer cells in research with mice and shows in green light the growth of bone tumors over time. GFP is widely used as a biological tracer to tag disease-causing microbes so it can be seen as it colonizes and spreads inside live animals. We have even used GFP labeling in our research. When we were working on a coral disease caused by the fungus *Aspergillus sydowii* (described in chapter 3), a colleague engineered a strain of the fungus that was tagged with GFP, so we could readily see our pathogen glowing green in a petri dish or inside the infected coral. In chapter 4, I describe how a colleague at Cornell uses GFP as a visual fluorescent marker to tag and then track the nematocysts of sea anemones through the digestive system of a nudibranch. GFP can thus be used to visualize specific cell types in living animals, organs, and tissues. This is widely used in fields such as immunology, neurobiology, development, and cancer biology.

While the chemical structure of GFP is no longer a puzzle, the mystery of how and why it is used in nature remains. In the crystal jelly, blue bioluminescence flashes when a release of calcium triggers the photoprotein aequorin. But sometimes the blue light is transduced to green flashes in a ring around the entire margin of the bell by interaction with GFP. No one knows any function for the eerie green ring of GFP that can only be activated by the blue flashes of aequorin.

A diverse bunch of ocean critters, from bacteria and jellyfish to squid and sharks, flash multicolored wavelengths of light in the dark. I have focused on cnidarian jellies because of the value of GFP and aequorin. Even though they have to steal some parts from their zooplankton prey, they use those parts in the synthesis of the giant bioluminescent chemical aequorin, with its spectacular calcium hair trigger.

As we continue to explore the unusual and useful powers of invertebrates, it is worth considering here that some of our prominent examples require early partnerships and trading powers across the tree of life. Sponges need the chemosynthetic ability of bacteria to build their potent, biologically active chemicals; squid need bioluminescent bacteria housed in pouches for countershading; corals need microalgae to build castles in low-nutrient waters; and jellies steal the building blocks of bioluminescence from their zooplankton prey. These partnerships show us an interconnected web of adaptations; we can't take even the smallest, simplest, or most alien-seeming creature for granted. Components of biodiversity we have yet to discover may be a necessary component to some creatures' greatest superpower. While on one stage we view the evolution of life as a fight for survival in a dog-eat-dog world, on another stage plays out the evolution of cooperative symbioses to pitch new superpowers in this game of life.

The diversity of sea star forms, from five armed to six armed to multiarmed. The ochre star tops the rock and the multiarmed sunflower star lurks in deeper water.

8.

The Sea Star's Sticky Skin

Sea stars prowl the rocks on intertidal shorelines, hunting for slow-moving prey like clams, oysters, and mussels. Unlike a panther, they aren't lightning quick or ferocious with big eyes, fangs, or claws. Instead, they move slowly on five arms with thousands of suckered tube feet. Unexpected as it seems given their strange body form so unlike the bilateral symmetry of vertebrates, sea stars emerge from deep evolutionary time as a powerful ecological force capable of transforming the seascape by virtue of their predatory prowess.

It baffles me that sea stars are such an underwater power, because they look improbable based on a more usual model of animal design. Most animals on land, whether vertebrates like cats and birds or invertebrates like insects, as well as many creatures in the ocean, have bilaterally symmetrical bodies like ours, with their brains and eyes up front. Sea stars have no visual attributes that would predict success as a predator; they have no brain, their eyes are simple photoreceptors on the ends of their arms, and their mouth is at the center of the often five radiating arms. They are pretty slow moving because their mobility is run on hundreds to thousands of tiny tube feet, a bit like a five-armed caterpillar or millipede. The tube feet are powered by a water hydraulic system that channels fluid through pipes and periodic gates, running through the stars like some ancient irrigation systems and culminating in fluid pressure controlled by muscles in each of the tiny tube feet. There are over two thousand sea star species in nature, and while most have five arms, some have more in multiples of five, and a few rare ones have broken the code to make six or eight arms. No matter how many arms they have, sea stars are also biologically strange, with multiple digestive glands and gonads, one in each arm. Sea stars have little ability to control their salt balance, so they can live only in the ocean and never in freshwater or on land. As frosting on the cake of strangeness, sea stars compete with octopuses for the most unusual skin in the animal kingdom. It can change to be almost instantly stiff or slack. The change is fast because it is under neural control; the electrical responsiveness is why it is called smart skin.

Smart skin, tube feet, water vascular system, and radial symmetry are all characteristics of the group of animals called echinoderms, meaning "spiny skin." Other relatives in this group are sea urchins, brittle stars, and sea cucumbers. They are all extremely ancient, originating over 540 million years ago, but strangely successful and very common in our modern oceans. They are also an extreme departure from all other invertebrates I have so far discussed. Echinoderms are more closely related to humans than they are to mollusks, cnidarians, or sponges. The key character that echinoderms and humans share is not obvious, because it is hidden in their early development. Our eggs divide in the same pattern, and our embryos' cells roll around and organize themselves in the same way. We are both deuterostomes, meaning that our mouths are second to develop, behind our back end. I think this means we are both back-ass-wards, at least compared with the mouth-first group, which includes snail and octopus relatives. After these common characters in development, the only other clues to our shared ancestry with echinoderms are similarities in our genes, which places us adjacent on genetically based ancestry trees. Let's now return to our rocky shores to consider the unusual story of how sea stars won fame in nature.

Ochre stars are so good at controlling populations of mussels that they play a key role in shaping the ecology of the rocky shoreline. Bob Paine, an ecologist from the University of Washington and one of my mentors, highlighted their mastery and founded an essential tenet of ecology after many years of research on remote Tatoosh Island. The star of his show was the vibrantly purple or orange ochre star, a species called *Pisaster ochraceus*, which can form a virtual rainbow on Pacific Northwest shores. Bob claimed this star had superpowers; it could control the entire functioning of a coastal seascape.

Tatoosh Island is almost in Canada, on the spectacular, wave-crashed, remote outer coast of the Olympic Peninsula. In this rugged seascape, Bob noticed that when ochre stars were absent, the mussel bed was like a vast black glacier that covered the rocks as far as he could see. When stars were plentiful, they ate buckets of mussels, and the entire mussel bed was denuded from the upper rocks and pushed into deeper water. This opened up space on the rocks for a multitude of other invertebrates like sea anemones, sponges, chitons, limpets, crabs, and lush algae. Could the slow-moving ochre star be the force behind this change?

To test his idea that a few stars were all-powerful in driving big change in the seascape, he went out one week and removed all of the stars from one section of shoreline. For comparison and to see what would happen, he added them back to an adjacent shore. Sure enough, after a few months, the stars had worked their magic. Where they were abundant, they ate all the mussels on the shallow edge, so the top edge of the mussel bed started much lower into the intertidal, making room for other critters. Where they were absent, the mussels spread out and covered all the rocks from the waterline to high up in the intertidal. Of course, this was just the big visible change. But Bob was brilliant in noting the profound functional change that went along with the visual: the areas without mussel beds had a much more diverse assemblage of other species.

An ecological keystone species is one in which a few individuals drive big ecological change. Bob turned these observations and experiments into a theory for how keystone consumers could structure any ecosystem. He mathematically defined a keystone consumer as one that has an outsize effect on prey, so a few stars remove enough mussels to cascade into significantly changing the seascape. Other examples of keystone consumers are lions that clear the Serengeti of wildebeests, otters that clear kelp beds of urchin hordes, and some predatory stream fish that transform rocky stream beds into algal meadows.

As a graduate student, I went along on the fishing boat with Bob's work crew to Tatoosh and got to see firsthand how this looked. Tatoosh Island is wild and excitingly remote, a pristine wilderness that is closed to almost all visitors. It is a half mile offshore from Cape Flattery, the most northwestern point in the continental United States. From Cape Flattery, you look straight out to the open Pacific Ocean and across the waves crashing on towering rock pillars to see the lighthouse atop the steep, rocky shores of Tatoosh Island. These are the remote, beautiful tribal lands of the Makah people. Forty years ago, we reached Tatoosh aboard a fishing boat, with permission granted to Bob by the Makah tribal leaders for his work. The crossing is never smooth at the best of times in these rough open-ocean waters. The approach to Tatoosh in the usual overcast weather is foreboding, with the sound and sight of enormous waves crashing on steep, rocky cliffs and sharp, rocky outcrops jutting from shore. It feels like the stormy British Isles murder mystery settings of Agatha Christie or Ken Follett. Landing on the island is a challenge; the docking zones are few and only possible in calm water at high tide. We came ashore by transferring to a rubber Zodiac skiff and then jumping ashore from a bouncing boat to a narrow, rocky point.

The tide was low at dawn the next morning, and I walked with Bob to help him weed stars from his starfish removal site. We followed the trail down from the lighthouse cabin and came out on a rocky bench. We stood and watched as huge waves crashed on the shore and frothed up to where we stood. Bob gave his rogue wave lecture: "As large as some of these waves look, there is always a bigger one coming. The rogue waves can be unexpected and their extent not obvious until they hit the shore. One of you must always be spotting for those working the intertidal. We have no way to retrieve you out here if you get pulled off the rocks. . . ." I've retold his lecture on the many intertidal shores of my career (Hawaii, San Juan Island, Appledore Island, Monterey) and heard countless stories of the people pulled off the intertidal while working there. His words and imagery from that day helped me keep my people safe.

The tide was still too high to get to our site. In the early morning light at low tide, the bands of sea creatures stood out as sharp strips of different life. Red and brown and green bands of turfy algae, shiny black mussels in a huge glacial bed, bright purple and orange starfish, and green anemones the size of dinner plates. Pale green sponges filled in the edges and under overhangs. As the tide slowly dropped, we walked along a surge channel, and Bob pointed to the section of rocky intertidal where he had removed sea stars years earlier.

The rocks were covered with a mussel bed of monumental proportions, reaching high up into the intertidal. Tatoosh is a special, high-biodiversity place, a plankton-rich site subsidized by nutrient upwelling and strong ocean waves. These were big mussels, grown as large as footballs, living attached to rocks with huge, plankton-filled waves to bring them unlimited food. The mussel bed crowded out all other creatures trying to find a space on the rock, like green anemones, green sponges, pinkish sea squirts, and diverse algae. Around the corner, in a surge channel where Bob had added ochre stars, the intertidal was bare of mussel bed and instead covered with brown, green, and red algae, bright pink coralline algae, green anemones, red sea squirts, and spiny purple urchins. This comparison demonstrated that the addition or removal of a few individuals of a single species, the ochre star, could control the entire appearance and function of a seascape like this intertidal. It's amazing that he even figured this out, because all you see at first is a few sea stars attached, unmoving, to the mussel bed. He reached the insight that these few stars created big change by watching them for years and realizing they did move around, ate a lot, and were actually a superpowered predator. Bob described the ochre star as a keystone species that functioned like the keystone in a stone arch: they hold up the arch of the intertidal biota, and the arch falls or changes without it. Bob coined the term "keystone species" in a 1969 note about food webs. The concept was quickly adopted and applied in other diverse ecosystems where small numbers of organisms from streams to the Serengeti could be big change agents on the landscape. It stands as a linchpin of ecology today and is widely used by wildlife and habitat mangers striving to return function to damaged ecosystems.

At six four, Bob was an outsize man and loomed a full foot taller than I. He also had an outsize personality and told outsize stories to support his theories. As a graduate student, I heard almost nonstop from Bob about the importance of running experiments in the field to test hypotheses about how nature works and show the importance of keystone species. I probably heard about the ochre star every week of my four years as a PhD student at the University of Washington. But exposure in my formative years to the theme of running experiments to test how predators could control an ecosystem had its effect on me. To this day, the ochre star is an all-powerful, mythical animal to me.

What is the secret power that puts sea stars on top in so many undersea habitats? Sea stars love clams, oysters, and mussels. These are all tricky to eat, impossible for most ocean predators to open by brute force. The rest of us, from humans to seagulls and crabs, simply break them in one way or another to get inside. Most people have not tried to open live clams, because we usually steam them before eating. But like me, you may have wrestled with opening a live oyster. You know that the only way in is to slip a sharp knife through the back of the shell to slice the adductor muscle, and then the oyster pops open. Despite your big muscles, it is impossible for you to open an oyster with brute strength. The sea star's superpower is the unprecedented strength and tenacity it brings to the task of opening its hard-shell prey. No other animal has the strength and persistence to pull open a clam, fighting against its impressively strong muscle. A clam does not open up right away, even when huge force is exerted. When touched, the clam clamps its shell shut with a tight adductor muscle. The next part is a battle of strength and time, and the sea star always wins with its constant, strong tension and intermittent pulls, using its hundreds of tube feet to grasp the clam and cling onto its shells. The suction created by the tiny tube feet is strong enough to hang on for hours under big forces generated by the prying arms of the star. The clam clamps and the sea star pulls, an arm wrestling match that can take half a day. When that force is applied for several hours, the adductor muscle inside the clam begins to fail, and the clam finally opens enough that the star can slip its enzyme-producing stomach inside the clam. The superpower to pry open the clam requires not only extremely strong, tenacious tube feet, but also a contribution from the strange skin of the sea star. The remarkable, alien properties of this skin are not a power scientists could have foreseen one hundred years ago.

The tricks a sea star uses to open a clam or mussel have been a mystery for at least the past hundred and fifty years. The two most likely possibilities considered initially were (1) the toxin theory, suggesting that sea stars secrete a substance that relaxes the adductor muscles of the bivalves; and (2) the mechanical theory, suggesting sea stars have enough strength to pull the valves of the bivalve apart with their tube feet.

Jacques Amand Eudes-Deslongchamps, a French surgeon and paleontologist, proposed the toxin hypothesis in 1826. As recently as 1932, Eisiro Sawano and Kinji Mitsugi, from Tohoku Imperial University, showed in experiments that extracts of sea star stomachs interfered with the beat of a mussel's heart, and proposed this as evidence supporting the toxin theory.

The other theory, that great strength could do the trick, was advanced originally by Paul Fischer, of the Paris National Museum of Natural History, in an 1864 report and later supported by Francis Jeffrey Bell, of the British Museum of Natural History, in his 1892 *Catalogue of the British Echinoderms*. Early experiments did show the tube feet were capable of very strong pull. In 1895, Paulus Schiemenz measured that the valves of the clam *Venus verrucosa* could be separated by a pull of nine hundred grams and that it took one thousand grams to pull a clam away from the grip of an *Asterias* star.

Marcel Lavoie tested both theories with the midsize sea star *Asterias forbesi* at the Marine Biological Lab in Woods Hole, Massachusetts, in 1932. He both injected into a star and dribbled into the water the extracts from the sea star stomach. He tested the strength required to pull open a mussel and found it was the same whether or not an extract was introduced. In the sixty mussels he tested, the degree of gaping was less for specimens exposed to stomach extracts than for those placed in seawater alone. The extracts of sea star stomachs did not help open the mussel. He rejected the toxin theory on this evidence.

Testing the mechanical strength theory was harder. Lavoie removed the adductor muscle from some mussels and replaced it with a spring bolted to the inside of the shells that could measure the force exerted to pull open a mussel. Then he put the guts back in the mussel so it would smell appealing to the star.

The results were crystal clear. Stars exerted pulls in excess of 1.8 pounds for several hours to get a clam open. The tube feet could hold the mussel, but it was the inexorable pulling and holding strength of the arms that pried the mussel open. Next, a surprisingly small gape of 0.1 millimeter was all that was needed for the sea star to slide its stomach inside the mussel, at which point toxins did play a role. Once the stomach could slide in and begin digesting the clam with enzymes, the clam was finished. Lavoie's experiments also showed that the pull by a sea star is not steady and sustained but rather intermittent over the many hours it took to open a mussel.

The unusual biology of sea star skin is the key to its great strength. It can stiffen in seconds and remain stiff for hours. Sea stars stiffen in a hunched posture with many tube feet and all five arms pulling to lever open a clam.

It costs them almost nothing in energy because it is accomplished by components reorganizing in the skin. Sea star skin is spiny with embedded hard skeletal struts integrated with collagen fibers that can instantly link the struts into a stiff framework. A stiff posture allows the star to hold the optimal prying pose for a long time and at very low energy cost. The real prize is that this change in stiffness is activated quickly by nerves and not more slowly by muscles. It is fast and not energy consuming.

The way the skin works is that nerves trigger collagen fibrils embedded in the skin of the star to instantly cross-link the larger fibers and calcareous struts, a bit like a zipper engaging its teeth to secure an opening. The skin instantly stiffens and is anchored by its internal skeleton of calcified spicules. Lots of invertebrates, from sponges to soft corals to sea squirts, have spicules in their skin, but sea stars are the only ones with stiffness triggered directly by neural and not muscular control. It is called mutable collagenous tissue (MCT) and is unique in the animal kingdom. *Mutable* means "to change"; *mutable collagen* refers to ability of the collagen itself to change by cross-linking. Greg Szulgit and Robert Shadwick investigated the mechanism behind this mutability by subjecting slices of the skin of sea cucumbers (which are close relative of sea stars and have the same properties) to a battery of mechanical tests. They found that alterations in stiffness are caused by the linking and unlinking of long, elastic collagenous fibers running through the tissue. When force is applied to the outer skin—a few gentle taps will do—the elastic collagen fibers link up, making a tough, resilient network that runs through the entire tissue, stiffening it.

The stunning part to me is that smart skin can quickly go both ways. Stiff skin helps to open clams; slack skin helps to autotomize, or drop their arms. Regeneration is common in sea stars; most can regrow an arm lost to a predator. That is why I sometimes find four-armed stars with a fifth small regenerating arm. The autotomy process, coming from "self-severing" in Greek, begins again with a nervous signal that this time reduces the stiffness of the skin and causes it to split apart, right at the margin of an arm. Eventually the skin draws back and heals over and the arm drops off. It can grow back in several months. Some species have taken it to the next level by dropping arms that propagate an entirely new star. While I see the actual catch skin as the superpower that underlies this ability, the regeneration capabilities of sea stars are essentially a superpower in their own right and are studied as a model for therapeutics that could aid humans in repairing damaged cells or tissues.

The smart skin of sea stars and their relatives is of interest to engineers because it has big applications. It is considered a smart material with unusual attributes: it can change stiffness reversibly back and forth, it can change almost instantly, and it is under neural and not muscular control. Nerve cells generate electrical signals to activate change, and this means the change can be also activated by a human-supplied electrical current. Smart materials, also called responsive materials by engineers, are new-generation materials that change their properties rapidly in response to the flip of a switch. Human-designed smart materials have one or more properties that can be changed by external triggers such as stress, moisture, electric or magnetic fields, light, temperature, pH, or chemicals. The smart skin of sea stars, with their programmed stiffness changes, is a useful model for constructing hybrid smart biomaterials. Nature got there millions of years before the engineers did.

While we do not need mutable collagenous tissue to open clams like a sea star, we do need smart materials that can change their stiffness instantly at the touch of a button and charge of electricity or change in temperature. In 2000, John Trotter and colleagues took the design of a smart material one step further by following up on the discovery of the way the collagen fibrils cross-link in a sea cucumber. They designed a hybrid biomaterial assembled from sea cucumber collagen fibrils interlaced with a synthetic matrix. The synthetic matrix allowed light or electricity to activate and either assemble or reverse fiber cross-links and thus change the stiffness of the material. These smart materials, patterned after echinoderm skin, can be used to treat tendons and ligaments following surgery. Improving surgery of tendons and ligaments, such as the very common ACL surgery that my son had following a soccer injury, is needed. Tendons are tricky to repair because there are large forces applied in holding these joints together in athletes, plus it's hard to find easy donor sites for making new ligaments. My son's ACL was repaired with a piece of his own hamstring autografted into his knee. This required first an invasive surgery to grab the ligament needed to repair the knee and then a second surgery to stitch it in place. Development of new synthetic smart materials for such reconstructions could be less invasive and also stronger.

Skin, tendon, and bone are all human tissues built of the protein collagen. Understanding the molecular mechanisms underlying change in tissue properties like stiffness can be applied to develop new biomaterials with planned properties. Collagen implants are widely used for wound healing, and the collagen used in implants grows more readily with electric current added to control the direction and speed of collagen growth. The ability of mutable collagenous tissue (MCB) to undergo large changes in stiffness without changing size and shape simply by increasing fiber cross-linking is the marvel. Added to that, it is under neural control and thus can be controlled electrically. There is keen interest in development of MCT as a marine-derived biological material for reconstructive surgery.

Interest in the potential biomedical uses of MCT has been steadily growing since Trotter first introduced the notion of MCT-inspired materials. One approach relies on directly observing how MCT works to generate new concepts in the design of materials that mimic its properties. This uses the organism's superpower as inspiration to imitate the properties in a human transplant. Another approach is to actually harvest an echinoderm tissue and apply that directly as a transplant to humans.

When the keystone is removed from an arch, the structural integrity of that arch is broken, and it falls. Could the same really be true for a keystone species in nature? Could the death of a single species with relatively small biomass truly cause fundamental ecosystem change? The key to the big effect of sea stars in Bob's experiments is the simple fact that each sea star eats a lot of mussels, and mussels are normally the dominant space holder in these intertidal habitats. So removing so many mussels creates space for a much wider diversity of other plants and animals. We were about to see a more widespread example of a keystone species effect in a removal experiment that far exceeded anything Bob Paine had ever imagined. In 2013, the ochre star began dying en masse on the West Coast from San Diego to our shores in Washington and north to Canada and Alaska. Next, a large deepwater species, the sunflower star (*Pycnopodia helianthoides*), started dropping dead by the millions in Canada and south all the way to central California. An additional seventeen species began dying catastrophically along the Pacific coast. The onset of this scale of mass mortality felt like a three-alarm fire. The cause of such widespread, rapid death was the mystery of our time.

My first sighting of the damage was at 8:00 p.m. in mid-December 2013 on a dark beach in full sight of the Seattle city lights. As I walked the dark shoreline at low tide, I saw sea star arms and parts of bodies from two species strewn all along the tideline, highlighted by my headlamp. It looked like all the ochre stars and a related intertidal species were dead and dying near Seattle that night. I knew this was also happening to the south in California and to the north in Canada. My brain raced ahead, and I had a vision of our stable ocean world collapsing and changing in what felt like an instant. Each star that was dying was a real being, with a history and a place in our world. Each star had begun life as a tiny larva and survived insane challenges to make it to this beach and metamorphose into a juvenile. Each star that was here had paced these shores as a juvenile in search of tiny baby mussels that were small enough for their little tube feet to open. Many stars had grown up on this beach for over twenty years to produce their own baby stars. And yet, here they were, massacred, reduced to bodies with ripped-apart arms. My legendary marine keystone crashed, and the entire ocean suddenly appeared so fragile. It was horrific, but somehow a metaphor for the larger change that was coming.

The mass mortality of these iconic species, beloved of children and adults alike, lighted a fire in everyone involved. A PBS newscast by the famed Gwen Ifill describing the devastation reached well over a million viewers in January 2014. Children and their parents saw dead sea star bodies and arms strewn across their West Coast beaches; children in the heartland who only dream of sea stars and beaches were also shocked. Most of my research is funded by government agencies, but this happened too quickly for normal funding. Instead, Mrs. Bailey's fourth-grade class in Arkansas, distraught at the reports of lost sea stars, launched a fundraising event, selling small cards and T-shirts that read SAVE A SEA STAR. One day, a check from them for $200 arrived in my mail. I realized how much just the existence of our stars meant to these kids and was moved by their determination to help. I did actually cry, touched by their effort, but I also decided we would put their funds to good use. So I matched their gift, and one of our donors matched and magnified both our gifts. This was enough to fund our surveys and early work to determine that it was an infectious microbe that was sickening and killing our stars.

Our project to understand the outbreak that has killed sea stars is also all about perseverance and holding on. After ten years, I am proud that in 2024, our team, led by Dr. Alyssa Gehman and funded by Hakai Institute and The Nature Conservancy, finally had the culprit in culture. Knowing the identity of the microbe will allow us to track it in nature and search for environmental reservoirs and refugia.

It's a tricky business watching to see how all twenty different sea star species fared. Although the ochre star was more common and easily studied in the intertidal, we took a particular interest in the sunflower star, because it was the most common in our deeper waters, was the quickest to die, and is a link in the chain of survival for our kelp forests. We chose it for our initial experiments because, of all the species, it picked up infections fast and died rapidly, thus it seemed the most susceptible and would be the most informative test animal. But we also knew little about it because it lived in deeper water, so it took longer to detect an outsize decline in nature. By the time we had enough data from citizen divers three years into the epidemic, this star had disappeared from California and was severely endangered all the way to Alaska. This study also pinpointed the role of climate warming in accelerating the speed and increasing the level of mortality. Thousands of diver surveys documented that the sunflower star is 97 percent absent from the continental US, and Dr. Sarah Gravem and Sara Hamilton, a researcher and graduate student from Oregon State University, respectively, compiled the data to support listing it as an endangered species with the International Union for Conservation of Nature. Along the way, a new truth emerged.

The sunflower star itself was previously unrecognized as a keystone species. Its widespread devastation, particularly in California, but also into Washington and British Columbia, triggered big ecological change; urchin populations exploded, and kelp beds collapsed. Removing sunflower stars creates this domino effect because they control populations of sea urchins, both by eating tons of tiny young urchins and also by stampeding away aggregations of adult sea urchins. When all the sunflower stars died rapidly, the urchin populations exploded and eradicated their food, the giant kelp forests. The loss of kelp meadows reduced important nursery grounds and habitat for commercial fish and crabs. To try to recover valuable kelp forests, we now work on a coast-wide sunflower star recovery effort, led by the Nature Conservancy, several aquariums, and state agencies and including over seventy scientists. Our program includes captive breeding, epidemiology, and plans to outplant healthy stars in hopes of renewing some endangered kelp meadows that have been destroyed by exploding numbers of hungry sea urchins. The previously unknown importance of this giant star in controlling one of our most valuable nearshore habitats, is like the game of Jenga. We never know when the unanticipated loss of a species can change the seascape. The undersea network remains a frontier with a lot of unknown territory and unidentified keystone species.

 A few years ago, we were keeping giant sunflower stars in our big outside tanks at Friday Harbor Labs. Before the outbreak, the sunflower star was as common underwater on the West Coast as a robin in a tree. Other than its susceptibility to an infectious disease, kind of like Superman and his susceptibility to kryptonite, it was king of the oceans by virtue of its unusual strength. Sunflower stars are the largest sea star on the planet, up to three feet across, with twenty-four arms. They are fast predators and eat almost anything they can catch in the wild. But their favorites are slower-moving prey like clams and oysters. In our tanks, they would lurk quietly, gripping along the sides with their twenty-four arms and fifteen thousand suction-like tube feet. Sometimes they would activate thousands of their tube feet and circle the tanks at a rapid pace of ten feet per minute. Think of a pacing tiger in a cage. Especially at feeding time. Yes, they knew when this was; when one of us approached the tanks to feed, they would gather upside down at the surface in a begging posture with outstretched tube feet.

Each of our sunflower stars had weird, individualized quirks of behavior, like being good at catching clams or being fast to line up at mealtime. Dr. Jason Hodin, who runs the captive breeding program at Friday Harbor Labs and keeps track of the fertility status and sex of his animals, names them. Sunny is a stunning orange-yellow girl, Prince has bright purple hues in his skin, Prospero is as manipulative as *The Tempest*'s duke, and Leo prowls his aquarium like a lion.

A clam thrown into a deep tank does not sink quite as fast as a stone. It has a certain amount of drag, and sinks in a slow spiral. Amazingly enough, the hungry sunflower stars could stretch out an arm, sticky with tube feet, and catch a clam before it hit the bottom.

The sea star's superpower then activates, stretching tens of tiny tube feet, each smaller than a ladybug, and grab the clam. Each tube foot is adhesive with a suction cup on its bottom, controlled by muscles inside. When the muscles contract, it pulls up the center part of the tube foot, and suction is created. So the sea star uses between ten and one hundred tube feet simultaneously to rapidly grab and hold the clam.

Work is still underway to understand the mysterious underwater disease epidemic that was strangely transmissible and affected such an unusually wide range of sea star species. As I detail in my book *Ocean Outbreak*, the knowledge gaps are much larger, and the funding for work on underwater outbreaks is much less than for human outbreaks. So it's not unusual for it to take a decade or more to unravel the causative agent in an underwater epidemic. This turned out to be a sort of COVID-19 for sea stars in terms of its impact on their populations, but so much less is known than the fantastically rapid progress made for a human disease.

In our tale of two sea stars, the story is different for the ochre star than the sunflower star. Impacts to the intertidal ochre star were well studied because the animals are easy to count when the tide goes out. We are still monitoring ochre star recovery in the San Juan Islands following the beginning of the 2013 pandemic. My favorite intertidal site to monitor is on the northeast shore of San Juan Island, a place that had 155 stars in our transects in 2013 before the pandemic. Recently, COVID-19 interfered with our counts. Again in 2022, I finally headed over to get a count, wondering how the ochre star had fared in our waters. To get there, I walked a long trail through an old-growth Douglas fir and cedar forest. It was mostly still and quiet, but some days I heard the hoots of a great horned owl and was spooked to realize it had been watching me, or the screeching of a hunting bald eagle, or the crash of startled deer, and often felt the silent eyes of the ever-present foxes. I wound my way down the trail to a pebbly pocket cove, sandwiched between two rocky headlands. Just as I arrived, I heard a swoosh and jogged more quickly to my rocky point to see the tall,

THE SEA STAR'S STICKY SKIN

dark dorsal fin of a big male orca surface just off the point, followed closely by a smaller-finned female and a tiny calf. The shore echoed with their loud exhalations. These were the Bigg's transient orcas, which had been hunting the seals along this shore for months. It felt like a good omen to see them just as I arrived. The rocks were slippery on this headland, and I picked my way carefully along, looking for and recording the size and color of ochre stars as I found them. Most were a deep, rich purple, but a few were orange, and one was pink. A few were really large, the length of a football, indicating they survived the epidemic and might be over twenty-five years old. All told, the good news is they were all healthy, but there were still only half as many as at the start of monitoring.

Although they initially also took a huge mortality hit and in some places the mussels surged as predicted, the ochre star has recovered to good levels in many places. We still see periodic episodes of wasting and ochre star mortality in warm periods of late summer and early fall, but there are also many survivors. The solid number of survivors is exciting and says clearly to me that some of our stars have evolved natural resistance to the infectious agent.

I settled beside a tide pool, watching a medium-sized purple star glide around on coordinated tube feet, and thought about the strangeness of this highly improbable, ancient invertebrate. What was happening in Precambrian seas to come up with this body plan? This one takes the cake when we talk about shaping the rules of biology. Its skin is as different from ours as an alien from another planet, with the strange property of instant stiffening and melting. Usually it reproduces sexually, but some of them reproduce asexually by dropping arms that completely regenerate a whole new body. It walks on hundreds to thousands of tiny tube feet that must all coordinate to hustle in one direction across rocky terrain. All this odd biology, and yet it comes out on top as a fiercely successful predator. What other secrets of biology lurk in this odd animal? The deadly sea star epidemic was an event on an epochal scale that left me thinking about long-term evolutionary processes and the changing biology of our ancient menagerie. Will this epidemic cause an evolutionary change in our sea stars that will persist for another hundred million years?

It's a good sign that many ochre stars are surviving and are likely more resistant to a deadly disease than the sunflower star. Evolution has been at its constant work of shaping new adaptations in a changing ocean. What secrets of invertebrate innate immunity will we uncover to explain how the ochre star immune system has changed and how much better its resistance to this disease is than the sunflower star's? Work is underway in our lab and others to understand the hidden world of the sea star immune system and whether we can pinpoint what elements, such as immune genes, are different in the stars that survive these events. While the biomedical applications of the strange skin of sea stars are startling and promising, I reckon we have other sea star discoveries ahead. We may find some unusual tricks in how this ancient immune system has evolved to become resistant to a fatal epidemic and discover new superpowers that promote a single unlikely species to a keystone ecological role in nature.

Epilogue

Spineless Futures in a Warming, Acidic Ocean

This ancient menagerie growing under the waves shows us new ways to think about rules shaping life and new solutions to better living. The most precious resource on our planet is not oil or metal, it is the deep secrets that string our web of life together. Nowhere is life's frontier greater or the potential riches more valuable than in our oceans.

In the 1980s, when I began diving and studying ocean life, it never occurred to me that the oceans might become less healthy and bountiful or inspire any emotions other than awe and love. And yet, here we are four decades later, the oceans under unprecedented pressures, their life so thinned and threatened that it is hard not to feel pangs of grief and worry for the future. I'm watching castles crumble and stars fall. An extraordinary menagerie with astounding value for humans and the stability of our planet is changing fast and disappearing, piece by piece.

We are living through the ever-increasing challenges that human-caused climate change detonates on land, undermining our survival with earth-rocking events like extreme warming, atmospheric rivers, mega-storms, rising seas, and lethal infectious diseases. The curveballs of climate change include insidious pitches like disease outbreaks that force us to up our game. When a previously unknown coronavirus emerged from the wild and overran and upended our globalized world, it reminded us that we know little about how changeable life is on this planet and how fragile our grip on survival is. Scientists not only tapped extraordinary knowledge of how to rapidly engineer newer vaccines, but also looked deeper to glimpse a remarkable diversity of coronaviruses infecting all our biodiversity, from clams to fish to bats. They noticed salmon with coronaviruses attacking the gills, reminiscent of the way our coronavirus attacked the human lung. I imagine that the diversity of life-forms infected with various coronaviruses is also an opportunity. Could studying these wild host-pathogen systems unveil new avenues designed by nature to resist disease that we could apply? I am that person who buys stock when the market crashes, so I look for hope, solutions, and opportunity even in catastrophe. What can we learn for human health from the ways that clams or salmon fight their coronavirus? We are truly all on one mothership earth, playing out science fiction scenarios when it comes to the strange shocks like a global pandemic that changing biology sends our way.

We easily observe the ravages of climate disasters on land, but what is happening underwater? The oceans, comprising 97 percent of the water on our planet and 70 percent of the surface area, have

SPINELESS FUTURES IN A WARMING OCEAN

played no part in emitting greenhouse gases and indeed help by absorbing a quarter of the carbon dioxide in our atmosphere, buffering us on land from even more extreme impacts. However, the oceans suffer the consequences of this extra carbon dioxide, which reacts chemically and increases the acidity of ocean waters. In fifty years, the oceans have become 30 percent more acidic on average, and this obscures the more extreme local acidification events in some places like the Pacific Northwest, where the pH in Puget Sound has decreased three times faster than the open ocean. The seasonally low-pH waters of Puget Sound and offshore dissolve the calcareous shells of plankton, leaving them pitted, thin, and fragile while also killing oyster larvae in hatcheries.

In addition to directly acidifying the oceans, the greenhouse gases in our atmosphere have warmed the entire planet, including the oceans. The world's oceans absorb excess heat, and they heat faster than the land. The oceans have warmed over 1.4 degrees Fahrenheit on average in the last one hundred years, but the bulk of the warming has been in the last forty years, and the real killer has been extreme localized increases called heat waves. The current rate of warming is unlike any experienced in the last 66 million years, as recently confirmed by the new Intergovernmental Panel on Climate Change (IPCC) report. The IPCC is a world panel of distinguished experts on climate change established by the United Nations in 1988. These changes are affecting our ocean biota, from destroying temperature-sensitive coral reefs, to dissolving plankton and oyster shells, to melting polar bear habitats, to changing patterns and speeds of great ocean currents.

My time underwater in the most beautiful ocean refuges has

crashed into the reality of climate change and loss of habitat compounded by other big stresses like overfishing, pollution arising from population growth, and overconsumption. It's beyond my scope here to detail the impact of all these threats, but it's important to acknowledge that all these factors are not only problems in themselves but add more impact by synergizing together. For example, the infectious microbes introduced by aquaculture or human sewage become even more explosive threats in waters warmed by climate and fed by pollutants that nurture microbial growth. My research team has been tasked at different times with recording mass mortality events affecting corals, sea stars, and seagrasses, all associated with warming events. I feel these deeply as both personal losses and losses for the stability of our planet.

Climate change has resulted in both winners and losers in our spineless managerie, but mostly losers. Ironically, some of the decline is abetted by the superpowers themselves. Although we can track decline and returns of big vertebrates like whales and turtles, it's harder to chart the course of our most ancient life-forms, because we have imperfect records of spineless life on our planet. Species like rare jellyfish, octopuses, or nudibranchs can slip away unrecorded, without a trace. But some invertebrate groups are well documented, and we can turn to these for insight into the welfare of ocean invertebrates.

The biggest victims of climate change have been corals, the mighty builders of reefs that house over 25 percent of the ocean's biodiversity. Our accurate estimates of coral reef change show the news is not good. Following the coral bleaching event that stretched around the globe in 1997 and 1998, coral reefs worldwide have suffered such extreme losses that it's hard to keep track. By 2008, following multiple catastrophic bleaching events, one third of all coral species were declared at risk of extinction in a worldwide survey by the International Union for the Conservation of Nature. This survey is now almost twenty years old, and the number of coral species at risk has certainly increased with successive warming events. The risk of extinction is higher for corals than for any other organism, plant or animal, on our planet.

While coral reefs the world over have been severely affected—like in our own Florida Keys in multiple years, including a 2023 heat wave exceeding 100 degrees Fahrenheit—the best documented impacts of recent warming are recorded from the Australian Great Barrier Reef, as I recounted in chapter 2. Between 1998 and the present, only 2 percent of those reefs escaped bleaching altogether, while 80 percent have now bleached severely, with high coral mortality. Dr. Terry Hughes reports that the long-term outlook for this 1,400-mile-long ecosystem, arguably the world's best managed coral reef, is very poor. While it is difficult to confirm extinction in the ocean, we have undoubtedly lost multiple species of reef-building corals from our planet just in the last decade. From my vantage point as a graduate student, eagerly plunging in to study the world's ocean biodiversity in the 1980s, I could never have imagined that something so massive, global, and catastrophic was bearing down on us; that climate change could demolish hundreds of miles of coral reef on a planetary scale; or that climate change could eradicate an entire ecosystem, the ancient stronghold of marine biodiversity, with the most species of any habitat on our planet, in the span of my career.

It is a strange irony that the tremendous superpower of corals, their partnership with a photosynthesizing microalga, is the very Achilles' heel that is endangering them in our over-hot climate. The microalgae are even closer to their upper temperature limit than the coral itself and so become stressed and expel from the coral during warming. Without their solar power, most reef-building corals will starve to death because they have evolved over time to be dependent on this symbiosis; it is as essential for their survival as our own digestive system. Thus, the photosynthetic partnership that has carried corals to prominence for hundreds of millions of years is now a primary agent of their demise in a changing climate.

Greenhouse gas emissions are throwing corals a double whammy. Along with the heat stress and loss of their symbionts caused by warming, ocean acidification is reducing their ability to calcify and build strong castles. Already corals are detected with thinner skeletons in some locations where effects of ocean acidification are greater.

Despite this grim report for reef-building corals, some coral relatives have prospered. The soft corals and gorgonians are among these. We watched a process of gorgonian resilience when we studied sea fans on Caribbean reefs. Starting in 1988, the hard corals began widespread bleaching and dying. While some of the gorgonian corals are photosynthetic and also bleached, others rode out the warming events. Our sea fans were among the interesting cases, surviving a widespread fungal epidemic. Our work showed that their immune responses spiked during warming events, and we think they eventually prevailed due to the evolution of strong immunity during this period. So, in the thirty years between 1988 and 2018, while reef-building corals bleached and died, sea fans and other gorgonians in the Caribbean actually increased. Many Caribbean coral reefs are now composed of dead reef builders, topped with living soft corals. It is a strange irony that the superpower that first drew us to the gorgonians, the pigment-studded immune system, may be responsible for their survival in an ocean with microbes unleashed.

Gorgonians have been more resilient to heat waves like the 2014–16 event that killed scleractinian corals. A new study that asks "What makes a winner?" shows another twist in a different gorgonian superpower prevailing during heat waves: the power to form symbiotic partnerships with dinoflagellate algae. Gorgonians not only have different immune systems from scleractinian corals, but also invest in different species of algal symbiont. Most Caribbean gorgonians host an algal symbiont called *Breviolum*, which turns out to be quite heat resistant and genetically variable, thus evolutionarily adaptable. While the origins of a partnership with an alga different from that hosted by scleractinian reef builders are shrouded in deep evolutionary time, it is an eerie reminder of the roles of chance and circumstance in evolutionary persistence. However these differences in algal partners initially evolved, the new insight of seeing better outcomes with different algal symbionts may give us a new tool in building more resilient reef builders.

Sponge populations have gone both ways, declining due to warming and increasing due to other human-induced changes. Some of the large barrel sponges of the Caribbean have succumbed to disease during warming events, but others have gradually increased in number. Barrel sponges increased in reefs in the Florida Keys on average by 46 percent and 33 percent from 2000 to 2006 in plots on two different reefs. With reductions in predatory fish and live corals, encrusting sponges have crept from crevices in the reef where we watched them in the mid-1980s, overgrown live and dead corals, and populated more widely. Sponges as a group have benefited from more nutrients in the water and warmer temperatures, and maybe even by more frequent, bigger storms, which increase the currents that bring food around them. What we do not know is how warming and pollution have affected their sponge superpower to conjure biologically active chemicals or whether these climate impacts are affecting their bacterial symbionts.

There have been news reports of jelly swarms clogging intakes of power plants and beach closures due to swarms of dangerous stinging jellies. You would think from these news reports that jellies have indeed risen to new prominence in the oceans, but a Blaschka-inspired watcher like me who cares also about the diversity of many different jelly species sees a more nuanced message. Indeed, a few species of so-called nuisance jellies have increased to spread in swarms, benefiting from lower numbers of fish predators and higher nutrients. The swarms are causing a lot of trouble. However, there are many jelly species that are more rare or may have disappeared in today's oceans, so I am not yet ready to proclaim that jellies are winners in the underwater climate change lottery.

The question of whether the jelly power of light generation is affected by climate change is intriguing. The most basic possibility is that some bioluminescent organisms might increase in warmer waters, and this in fact has happened. Bioluminescent glows off the coast of India occur when warming waters cause light-emitting phytoplankton to increase. Warmer temperatures and acidic conditions also push some reactions faster, so it is entirely possible that a warmer, more acidic ocean will be brighter underwater. While there are no data showing how warming affects bioluminescence, studies are starting to show that ocean acidity affects light production. Under current emission projections, average ocean pH is expected to drop from 8.1 to 7.7 by the year 2100. Scientists have shown that coral relatives like the sea pansy increase light production in more acidic seas, while others, like firefly squid, lose the ability to bioluminesce in more acidic seas.

It is challenging to monitor populations of small mobile animals like bright sea slugs. Studies show that in California, nudibranchs increase during warming El Niño events and decline in colder La Niña events. This boom and bust of adult populations seems largely driven by conditions that affect larval stage growth in plankton. El Niño causes warming and different currents that bring more larvae close to shore, where they can transform into adults.

The extent of nudibranch population changes is largely unknown, but I am concerned that habitat loss, pollution, and steep declines in their anemone and coral prey could fuel future losses of nudibranch species.

Octopuses and some of their relatives are considered big winners in today's oceans, largely because predatory fish that normally eat octopuses and squid have declined due to overfishing. As squid and octopuses have increased and backboned fish have declined over the last two decades, many coastal fisheries have switched from a focus on backboned fish to spineless squid and octopuses.

When I look at Cornell's collection of Blaschka cephalopod diversity and some of the fascinating, more rare deepwater species, I worry that while common bottom and mid-water species are increasing, we may be losing some less well-known versions of the deeper-water forms. We do know that the numbers of common octopuses in the water have changed with climate, predominantly with warming of ocean waters. Studies show that octopuses have smaller populations in warmer conditions. The long-term effect of climate change has led to smaller and smaller populations of common octopus.

Giant clams have been endangered by overharvesting in the Pacific for over fifty years, but even in protected refuges, the survivors are now dying of heat stress in warming events and of disruption of their obligate symbiosis with microalgae.

Rapid global climate change now poses major new threats to giant clams. Like corals, giant clams use their photosynthesized energy from zooxanthellae to reach enormous sizes. Warming directly disrupts their superpower to photosynthesize by stressing the microalgal symbionts. When stressed, these clams can expel their symbiotic microalgae and turn white. They bleach and then die, just like corals. Climate change also acts indirectly to diminish giant clams because building and maintaining strong shells is slower and more costly in an acidic ocean. Greater acidification itself causes higher mortality in the delicate juvenile stages of many bivalves, including oysters and giant clams. Due to declines, four giant clam species are currently listed on the Red List of Threatened Species kept by the International Union for Conservation of Nature.

Thus, as with reef corals, the very superpower of photosynthesis that carried giant clams to prominence for hundreds of millions of years and was once a strength is now a vulnerability in a changed climate.

There are repeated examples of climate change fueling more epidemics on land and under the sea, as I describe in my book *Ocean Outbreak*. The epidemic of sea star wasting disease is a poster child for how these impacts can undermine both biodiversity and ecosystem stability. It's been a gut punch to me that we lost a big component of sea star biodiversity on the US West Coast, starting in 2013. The decline of sunflower stars from the epidemic created a domino effect of decline in these ecosystems. Demise of these top predators triggered an explosion of herbivorous sea urchins, which then mowed down great rainforests of kelp. Thus, wide swaths of an essential habitat for commercially important crabs and juvenile fish have disappeared.

There are still many questions about the sea star outbreak, but data show that the disease impact was fueled by warming. It is also another case where the very superpower that causes success is linked with their decline in a changed environment. In this case, the collagenous catch skin of sea stars is one of the first tissues to fail in a sick star. Slack skin and rapid loss of arms are hallmarks of this disease.

The increasing threats to our ocean have transformed global marine research themes. In my case, I've turned from being an invertebrate biologist and ecologist trying to understand the basic ecological rules of underwater life to a conservationist searching for ways to sustainably manage and conserve ocean biota under stress. I retain my conviction that in addition to the wonders of technology, it is the wonders of biology and the value of nature that will create new solutions for a planet at risk. Technology can accelerate and amplify and add tools to our kit, but it's the basic biological discoveries that will drive transformative innovations. Protecting the oceans and their ancient menagerie is essential for our own survival on this blue planet.

We cannot in the next fifty to one hundred years stop the juggernaut of climate change we have started, but if we dial back and reach net-zero emissions, there will be fewer years of "warming in the pipeline." One thing we can do is double down to manage resilience in our oceans and watch for and even manage the new tricks of invertebrates and their superpowers in a changing world. The sheer diversity and great age of this multitude of unusual body plans offers some spectacular opportunities for new dimensions in the rules of life. We can count on evolution to act on these groups to show us new ways to adapt to change. I saw one example of a way we can harness evolution's innovative power in a changing ocean in 2023 on a remote coral reef in Fiji.

I have a new job that brings me into the realm of ocean policy to protect our oceans and its living biota. My job as science envoy for ocean protections with the US State Department is to build bridges, develop partnerships, share knowledge, and network to shore up protections for valuable ocean resources worldwide. My focus is supporting the global implementation of one of our most valuable tools in biodiversity management, marine protected areas (MPAs). The concept is simple: protecting valuable habitats like coral reefs, mangroves, and seagrass meadows from destruction, extractive fishing, and mining allows resident biota and associated services to thrive. While we often think about MPAs as one approach to preventing overfishing, another huge benefit is that an MPA preserves not just fish but a portfolio of valuable biodiversity and habitats for the future. Protected no-fish reserves nurture, produce, and export increased fish and invertebrate biodiversity that can spill over to surrounding locales. Our research team is taking a One Health perspective. One Health is an initiative that recognizes that the health of people is closely linked to the health of animals and our shared environment. I described in chapter 2 that not only do protected areas allow fish and habitat to thrive, but the corals inside protected reefs are healthier, with fewer diseases. We've also shown in two studies that the health benefits of lower pathogens in seagrass meadows reach to humans by reducing their contact with waterborne pathogens and increasing the safety of food from the sea.

While some coastal MPAs clearly confer greater health, less extractive damage, and a more intact ecosystem, can MPAs protect against existential threats like climate impacts, such as warming and ocean acidification? Can we take MPAs to the next level and think of them as engines of evolutionary innovation? On a very small, local scale, underwater plants like mangroves and seagrasses can reduce ocean acidification, because these plants absorb and bind up carbon dioxide. Other studies show that MPAs can reduce impacts of heat waves if the biota inside the MPA is healthier and able to better withstand stress or recover more completely after a heat event. One nice example of this is the faster and greater recovery of abalone in reserves following heat waves in Mexico. Pink and green abalone died en masse in heat waves, but populations recovered more rapidly inside reserves. This shows the value of combining resilience strategies, including combining climate refugia and marine reserves, adherence to conservative annual fishing quotas, fishing closures, minimum size limitations, and ecological monitoring. So a new challenge is to refine our approaches to adaptive management to contend with protecting a changing ocean. In the US, the Biden administration adopted a policy of thirty by thirty, which means to protect 30 percent of our lands and oceans by 2030. This policy recognizes both the climate mitigation value of nature and its other values. This policy is also adopted globally by the UN and is causing a tidal wave worldwide in developing more ocean protections.

What kinds of powers in nature are expected to benefit from more protections? I have focused in this book on the portfolio of superpowers of particular groups of marine invertebrates. But some oceanic habitats themselves punch above their weight in providing solutions to aid biodiversity and even human health. How can we not only protect these services but also use science and technology to help evolution keep up with threats like climate change?

On a recent trip to several Pacific Islands to help develop strategies for marine spatial planning, I visited an innovative project that combined protected areas with new approaches to helping evolution adapt more rapidly. Dr. Victor Bonito works with local communities in Fiji to create no-fish reserves. Inside those safe havens, he improves the coral biodiversity and resilience by selectively propagating the coral genotypes and species that best survive bleaching events. He is essentially assisting evolution by protecting, selecting, and propagating heat-resistant corals. What astounded me is that I saw how a single branch of coral can grow into a multi-branch colony ready for transplant onto the reef within ten months. Multiply this by several hundred transplants, and a functional reef can be restored in two years. Victor's project sounds like a simple coral-gardening job, but the innovation comes in with his scientific method, organized approach, and coral taxonomic expertise to select the right strains and species that withstand heat events. This is a superb model of managing protected habitats to be engines of evolutionary innovation.

In the fight to sustain coral reefs against the impossible odds of the climate warming juggernaut, scientists look to the coral's powers of regeneration and building and the possibility for evolutionary adaptation in the hope that some species have unusual capacity to evolve new strategies to withstand warming and the fresh onslaught of heat-fueled microorganisms. In addition to whole ecosystem field selection for heat tolerance like the work that Victor Bonito and Stanford professor Steve Palumbi and many others undertake, lab studies are engineering new algal and bacterial strains that confer more heat resistance. There are three main components that can be engineered to aid a coral's adaptation to rapid temperature change—the coral animal, their algal symbionts, and the bacteria on their surface—and scientists are engineering new approaches with all three in trying to increase heat tolerance of the coral holobiont. Humans and corals are now both in an existential race to adapt to the climate change sweeping our planet. Researchers the world over are pushing the envelope to aid coral adaptation to extreme heat waves.

This same example to speed evolutionary adaptation to a warming climate could be applied to future proofing many ocean habitats containing ancient superpowers. For example, consider the value of protecting and adaptively managing seagrass meadows.

There may be no habitat on earth with greater superpowers for health than seagrass meadows. Seagrasses are flowering plants that grow as vast meadows in shallow nearshore waters on both tropical and temperate coasts. They are a frontier of extraordinary processes for life. In addition to the well-known role of seagrass meadows as nursery habitat for fish and shore protections for humans, seagrasses have three newly recognized services in processing seawater: (1) seagrasses absorb excess nutrients and organic pollutants and thus clean water; (2) seagrasses absorb and bind up excess carbon dioxide and thus reduce atmospheric and oceanic carbon dioxide (called blue carbon); and (3) seagrass meadows detoxify pathogenic microbes in the water and thus reduce pathogenic bacteria for humans and other marine biota like corals. Our work described in chapter 2 showed that pathogenic bacteria in Indonesia and even in the urban seagrass meadows of Puget Sound are reduced substantially. This hygiene service of seagrasses is a new frontier showing previously unknown life-giving properties of natural ecosystems like seagrasses.

But, like coral reefs, seagrass meadows themselves are decimated by a combination of disease and warming. Could the assisted evolution approaches being developed for coral reefs help seagrass meadows adapt to a warming climate? I think seagrass managers will soon be testing multiple genetic strains for resistance to disease and heat stress in efforts to keep these valuable meadows afloat.

In addition to propagating new strains and preserving seagrasses in urban or polluted environments, discovering how the hygiene service acts in a seagrass meadow can lead us to better design resilient shore protections in urban areas. By thinking ahead to assisting evolution, we can also ask if certain strains of seagrass are more tolerant to heat stress and provide a larger hygiene service. Of course, we also need sewage treatment plants, but seagrass meadows have special services to contribute, especially during the projected increase of flooding and rain overflow of sewage systems. Seagrass meadows have unusual powers, because they rapidly destroy dangerous bacteria, operate 24-7 for free, adapt to and grow with the environment as it changes, and provide co-benefits like habitat and wave surge protection.

Can we use the portfolio of nature's services and superpowers to stem the tide of biodiversity loss in our oceans? There are local solutions, from protecting the seagrass or coral reef in your backyard to assisting the evolution of resilience in reserves, to all-encompassing strategies that reduce fishing or plastic pollution in the global ocean.

Big changes are underway that will elevate understanding and protection of our menagerie's superpowers and the value of the ocean's services for humans. In a new program, the United States is quantifying the value of nature. The goal of the National Nature Assessment is to systematically discover and log the value of our nation's natural resources; consider their climate and service value, distribution, and extent; and enact laws to protect the most valuable. Knowledge of the special superpowers in groups of marine biota will be essential data in this process. Heather Tallis of the US Office of Science and Technology Policy runs this program and says, "Nature-based solutions should be go-to options for addressing challenges like climate change and building strong economies." Todd Bridges of the US Army Corps of Engineers notes that the timing and the opportunity for this report's recommendations are unprecedented:

> We're in a time of important change or evolution in people's thinking about nature. Rather than predominantly thinking about nature as a source of threat, or something to be conquered or overcome, there's a growing awareness that nature and its ecosystems are a foundation, the source and supply, the economy if you will, to address these social and environmental challenges. For agencies like the Corps of Engineers who are charged with delivering service or projects of a particular type, I think the future is what I might call "nature-first thinking" where we think about how can nature help before we decide what action we need to take.

It is a new age for managing our oceans, with new technology for satellite remote sensing of temperature and other metrics, underwater imaging and drones, and astounding advances in molecular biology. The most spectacular new technologies in molecular

SPINELESS FUTURES IN A WARMING OCEAN

biology include environmental DNA (eDNA) to detect the presence of absent species that previously passed through the water and the ability to use gene editing to explore fundamental biological processes. Gene editing will create new ways to use some superpowers and engineer new ways to assist adaptive evolutionary change.

It seems like a big jump to talk about engineering biology when there is still so much to learn about how the superpowers of the spineless work biologically. We can see the functional change, for example, when a coral reef evolves a new heat-tolerant genetic strain, but we can't visually see the genetic material and process that drives it. CRISPR/Cas9 is a gene editing system that has revolutionized biology over the last decade by allowing scientists to edit functional genes. Scientists have utilized CRISPR/Cas9 editing to probe biological function, dissect genetic interactions, and inform strategies to combat human diseases and engineer crops. Amid a flurry of biomedical applications targeting human disease are the new CRISPR-based products—for example, more nutritious tomatoes and disease-resistant crops. A future I see for gene editing is to unlock genetic mysteries of invertebrate superpowers and even allow mixing and matching powers among the groups. Consider that one very special power of the invertebrates is their excellence in creating partnerships with bacteria. Consider also that CRISPR/Cas9 evolved as a bacterial editing system. What job do you think drove its evolution as an editor in bacteria?

It was discovered in 2007 that the biological function of CRISPR/Cas9 editing in bacteria is to improve immunity against viruses. The original evolutionary benefit of CRISPR is the ability for bacteria to rapidly cut and splice their genome to become more resistant to disease. In just a decade, scientists have gone wild using CRISPR/Cas9 technology for lots of other changes in organisms; they call it molecular scissors, developed by bacteria to trim and move DNA sequences around. It is now widely used to snip and replace genes in diverse species ranging from microbes and plants to animals and even humans, thus creating what we call transgenic organisms. In the extreme, think of dogs with functioning pig hearts.

The vexing ethical issues for humans aside, I see phenomenal promise in gene editing as a discovery tool in our spineless menagerie. Among the invertebrates, the greatest focus of CRISPR has been on engineering changes to insect pests, and this has advanced rapidly in a model species like the fruit fly, which can be easily manipulated genetically. The future is here with the production of designer flies. Many of the applications in insects are knockout gene changes, which explore how removing genes can change outcomes. If we knock out this gene, how will an outcome change? One example is removing genes essential for mating in pest insects like mosquitoes so that the mosquitoes cannot breed. The next level is to replace one gene with another, such as a transgenic vector-control strategy that converts female mosquitoes into harmless males. This vector-management strategy works well because only female mosquitoes feed on blood and transmit pathogens. So getting rid of female mosquitoes is likely good.

Another example of a useful CRISPR edit is the introduction of fungal resistance into honeybees, which are highly threatened with fungal disease. We can certainly imagine how useful genetic engineering could be to eventually increase our sea star resistance to its pathogen, and our lab has early work underway to understand if there are target genes to be edited into susceptible sea stars. Or maybe we can engineer and propagate more heat-resistant symbiotic algae for reef-building corals, building on what we know of different species of heat-resistant symbionts in gorgonians. I talked earlier about how CRISPR is being used to understand how nudibranchs steal nematocysts. Even bigger edits are afoot in medicine, and they link with what we know about the way some of our invertebrate superpowers are controlled.

For example, once any biological change can be controlled electrically through a nervous system—think of octopus or sea star "smart skins" or a beating heart triggered with an electrical pacemaker—it is possible to engineer in different triggers other than electricity, like light or temperature. For example, the heartbeats of pygmy squid, zebra fish, and mice are currently being engineered to be switched on and off with changing wavelengths of laser light. This is accomplished by splicing a gene for light sensitivity from proteins called channelrhodopsin and halorhodopsin, which are taken from viruses, into squid or zebra fish neurons that control heartbeats. The vision for the future is that a stopped human heart could be reactivated with laser pacemakers instead of jolts of electricity.

The future is here, with geneticists and CRISPR technologists redesigning the rules of life, to activate a beating heart with both electricity and light. Unusual biologies, like those of our most ancient menagerie, have new value providing new powers that can be designed into humans or medicines or agricultural products. I look forward to seeing more transformations in biology arising from our ancient menagerie and hope we can find the wisdom to protect our living oceans, which will continue to source our most fundamental new discoveries.

The spineless denizens of the ocean have taught me that the rules of life can be reshaped in this lawless enterprise of evolution. On the surface, it seems as though anything goes as biology finds strange ways around the making and preserving of life. Some of the most novel solutions are those born of strange partnerships across the tree of life; some of the most unusual of these originated in deep time in the oceans. The two most ancient protagonists in these partnerships are algae and bacteria. Algae and corals team to photosynthesize and make almost immortal living castles; algae and giant clams team to build big, long-lived clams; algae and sponges team to build giant barrel sponges. Bacteria team with sponges to craft spectacular chemicals with novel effects, and bacteria team with squid to produce flashing lights. Likely other key bacterial symbioses exist that are yet undescribed. Novelty on our planet exists everywhere in nature, but nowhere in such diversity as in the oceans. The deeper we dig into biological processes of spineless creatures and their strange partnerships with other entities like bacteria, the more a new frontier of discovery unfolds.

Acknowledgments

A book like this takes a village; all the kind, brilliant, inspiring help along the way has made this book a delight for me. One maker in this village is my book assistant, Audrey Vinton. She helped and encouraged me every step of the way, from attentive, smart fact checking on early chapters, to ideas and content on cephalopods, to the really arduous job of finding, checking, and putting into format every reference in the book. The other maker in this village is Eric Engles, from Edit Craft Editorial, who helped me with early design and edits on every chapter.

I am very grateful to my son, writer Nathan Greene, for editorial help, ideas, and discussions on very early versions of the book and thoughtful discussions about the title. I am similarly grateful to my husband, Charles Greene, for often engaging on short notice with the constant writing discussions and often last-minute decisions on emphasis and titles and content.

ACKNOWLEDGMENTS

Katherine Flynn, my smart, attentive agent, was enthusiastic from the start and helped guide the initial proposal and weighed in at various stages as needed throughout the book process. I always felt she was there for me. Emily Wunderlich, my talented brilliant editor at Viking, believed in the book from the start and patiently met monthly on Zoom to talk through ideas and content as chapters rolled out. Her consistent creative input helped set a steady course. She in turn unlocked an amazing team at Viking, including production editor Jennifer Tait, copyeditor Angelina Krahn, editorial assistant Carlos Zayas-Pons, publicist Julia Rickard, and marketer Anna Brill.

Three people read the entire book draft and commented helpfully: Eric Engles, Rick Grosberg, and David O. Brown. David kindly allowed me to use his beautiful photos. Steve Palumbi was always happy to chat about ideas and writing and gave input on chapter 4 and the epilogue. Vicki Pearse read an early proposal and the chapter 1 and very kindly allowed me to use illustrations from *Living Invertebrates* and *Animals Without Backbones*. Amanda Moon of Moon and Co. helped with early ideas.

It's been a dream come true to work with the fabulous invertebrate artist Andrea Dingeldein in this spineless endeavor. She patiently created the chapter diversity illustrations to fit each chapter's cast of characters.

I am so grateful for the community of friends, family, and expert colleagues who read and commented on chapters, some on more than one: Audrey Vinton, Olivia Graham, Chuck Greene, Nathan Greene, Morgan Greene, Clairie Ng, Kaite Cisz, Katie Lee, Michael LaBarbara, Laura Mydlarz, Richard Strathmann, Rita Pampanin,

ACKNOWLEDGMENTS

Eric Sanford, George Matsumoto, Steve Haddock, Dana Staaf, Leslie Babonis, Vicki Pearse, Steve Palumbi, Rick Grosberg, Julia Parrish, Nelson Hairston, Rachel Merz, Madelyn Hollister.

I appreciate Cornell University and my department of Ecology and Evolutionary Biology for over thirty-five years of moral support and intellectual inspiration. I appreciate Friday Harbor Labs and all the amazing invertebrates flourishing in so many healthy sea water tanks.

Notes

PREFACE

xi **In 2014, after** *The New York Times*: C. Drew Harvell, "Diving into the Coral Triangle," *New York Times*, January 29, 2013, archive.nytimes.com/scientistatwork.blogs.nytimes.com/2013/01/29/diving-into-the-coral-triangle.

C. Drew Harvell, "Living Reefs, Under Fire," *New York Times*, February 1, 2013, archive.nytimes.com/scientistatwork.blogs.nytimes.com/2013/02/01/living-reefs-under-fire.

C. Drew Harvell, "Taking Care of Eco-Business," *New York Times*, February 12, 2013, archive.nytimes.com/scientistatwork.blogs.nytimes.com/2013/02/12/taking-care-of-eco-business.

C. Drew Harvell, "Swift Diving in Bali," *New York Times*, February 19, 2013, archive.nytimes.com/scientistatwork.blogs.nytimes.com/2013/02/19/swift-diving-in-bali.

C. Drew Harvell, "In Pursuit of an Underwater Menagerie," *New York Times*, May, 6, 2013, nytimes.com/2013/05/07/science/blaschka-glass-menagerie-inspires-marine-expedition.html.

xii **It was published in 2016:** Drew Harvell, *A Sea of Glass: Searching for the Blaschkas' Fragile Legacy in an Ocean at Risk* (Oakland: University of California Press, 2016).

NOTES

xii **That book, *Ocean Outbreak*:** Drew Harvell, *Ocean Outbreak: Confronting the Rising Tide of Marine Disease* (Oakland: University of California Press, 2019).

xv **These smallest beginnings:** Michael G. Hadfield and Valerie J. Paul, "Natural Chemical Cues for Settlement and Metamorphosis of Marine-Invertebrate Larvae," in *Marine Chemical Ecology*, ed. James B. McClintock and Bill J. Baker (Boca Raton, FL: CRC Press, 2001), 431–61.

Michael J. Kingsford, Jeffrey M. Leis, Alan Shanks, Kenyon C. Lindeman, Steven G. Morgan, and Jesús Pineda, "Sensory Environments, Larval Abilities and Local Self-Recruitment," in "Open vs. Closed Marine Populations: Synthesis and Analysis of the Evidence," *Bulletin of Marine Science* 70, supp. 1 (January 2002): 309–40.

xv **My entire PhD thesis:** C. Drew Harvell, "Predator-Induced Defense in a Marine Bryozoan," *Science* 224, no. 4655 (June 1984): 1357–59, doi.org/10.1126/science.224.4655.1357.

xvi **It looks a lot like:** Andrej Ernst and Peter Königshof, "The Role of Bryozoans in Fossil Reefs—an Example from the Middle Devonian of the Western Sahara," *Facies* 54 (June 2008): 613–20, doi.org/10.1007/s10347-008-0149-1.

xvii **Carl Linnaeus created:** Carl Linnaeus, *Systema naturae per regna tria naturae, secundum classes, ordines, genera, species, cum characteribus, differentiis, synonymis, locis. Tomus I. Editio decima, reformata* (Stockholm, 1758).

xviii **In 1874, Ernst Haeckel:** Ernst Haeckel, *Anthropogenie oder Entwickelungsgeschichte des Menschen* (Leipzig: Wilhelm Engelmann, 1874).

xviii **The biological classification of animals:** Tetyana Nosenko, Fabian Schreiber, Maja Adamska, Marcin Adamski, Michael Eitel, Jörg Hammel, Manuel Maldonado, et al., "Deep Metazoan Phylogeny: When Different Genes Tell Different Stories," *Molecular Phylogenetics and Evolution* 67, no. 1 (April 2013): 223–33, doi.org10.1016/j.ympev.2013.01.010.

xix **And it is in considering:** Sven Beer, Mats Björk, and John Beardall, "Photosymbionts" in *Photosynthesis in the Marine Environment* (Ames, IA: Wiley Blackwell, 2014), 23–26.

NOTES

Dayana Yahalomi, Stephen D. Atkinson, Moran Neuhof, E. Sally Chang, Hervé Philippe, Paulyn Cartwright, Jerri L. Bartholomew, and Dorothée Huchon, "A Cnidarian Parasite of Salmon (Myxozoa: *Henneguya*) Lacks a Mitochondrial Genome," *PNAS* 117, no. 10 (March 2020): 5358–65, doi.org/10.1073/pnas.1909907117.

Jonathan D. Allen, Adam M. Reitzel, and William Jaeckle, "Asexual Reproduction of Marine Invertebrate Embryos and Larvae," in *Evolutionary Ecology of Marine Invertebrate Larvae*, ed. Tyler Carrier, Adam Reitzel, and Andreas Heyland (Oxford: Oxford University Press, 2018), 67–81.

xix **The best known of these:** Jennifer J. Wernegreen, "Endosymbiosis," *Current Biology* 22, no. 24 (July 2012): R555–61, doi.org/10.1016/j.cub.2012.06.010.

CHAPTER 1: THE SPONGE'S PHARMACOPOEIA

2 **A fossil recently found:** Zongjun Yin, Maoyan Zhu, Eric H. Davidson, David J. Bottjer, Fangchen Zhao, and Paul Tafforeau, "Sponge Grade Body Fossil with Cellular Resolution Dating 60 Myr before the Cambrian," *PNAS* 112, no. 12 (March 2015): E1453–60, doi.org/10.1073/pnas.1414577112.

3 **An ancient sponge lineage:** Françoise Debrenne, "Diversification of Archaeocyatha," in *Origin and Early Evolution of the Metazoa*, ed. Jere H. Lipps and Philip W. Signor (Boston: Springer US, 1992), 425–43, doi.org/10.1007/978-1-4899-2427-8_13.

3 **Living in shallow, sunlit waters:** George D. Stanley Jr. and Jere H. Lipps, "Photosymbiosis: The Driving Force for Reef Success and Failure," *Paleontological Society Papers* 17 (October 2011): 33–59, doi.org/10.1017/S1089332600002436.

Donna M. Surge, Michael Savarese, J. Robert Dodd, and Kyger C. Lohmann, "Carbon Isotopic Evidence for Photosynthesis in Early Cambrian Oceans," *Geology* 25, no. 6 (June 1997): 503–6, doi.org/10.1130/0091-7613(1997)025<0503:CIEFPI>2.3.CO;2.

NOTES

3 **The lack of mobility:** Joseph R. Pawlik, Brian Chanas, Robert J. Toonen, and William Fenical, "Defenses of Caribbean Sponges against Predatory Reef Fish. I. Chemical Deterrency," *Marine Ecology Progress Series* 127 (November 1995): 183–94, doi.org/10.3354/meps127183.

John W. Blunt, Brent R. Copp, Murray H. G. Munro, Peter T. Northcote, and Michèle R. Prinsep, "Marine Natural Products," *Natural Product Reports* 20, no. 1 (2003): 14–30, doi.org/10.1039/B207130B.

Rochelle W. Newbold, Paul R. Jensen, William Fenical, and Joseph R. Pawlik, "Antimicrobial Activity of Caribbean Sponge Extracts," *Aquatic Microbial Ecology* 19, no. 3 (October 1999): 279–84, doi.org/10.3354/ame019279.

9 **But over the last several:** Rajeev Kumar Jha and Xu Zi-rong, "Biomedical Compounds from Marine Organisms," *Marine Drugs* 2, no. 3 (August 2004): 123–46, doi.org/10.3390/md203123.

12 **These multipurpose cells can also:** Werner E. G. Müller and Isabel M. Müller, "Origin of the Metazoan Immune System: Identification of the Molecules and Their Functions in Sponges," *Integrative and Comparative Biology* 43, no. 2 (April 2003): 281–92, doi.org/10.1093/icb/43.2.281.

12 **By comparison, human cells are:** Christopher L. Baker and Martin F. Pera, "Capturing Totipotent Stem Cells," *Cell Stem Cell* 22, no. 1 (January 2018): 25–34, doi.org/10.1016/j.stem.2017.12.011.

14 **About a decade later, scientists:** Fehmida Bibi, Muhammad Faheem, Esam I. Azhar, Muhammad Yasir, Sana A. Alvi, Mohammad A. Kamal, Ikram Ullah, and Muhammad I. Naseer, "Bacteria from Marine Sponges: A Source of New Drugs," *Current Drug Metabolism* 18, no. 1 (2017): 11–15, doi.org/10.2174/1389200217666161013090610.

14 **I think I am correct:** D. John Faulkner, Mia D. Unson, and Carole A. Bewley, "The Chemistry of Some Sponges and Their Symbionts," *Pure and Applied Chemistry* 66, nos. 10/11 (1994): 1983–990, doi.org/10.1351/pac199466101983.

14 **Both bacterial strains were isolated:** E. W. Schmidt, A. Y. Obraztsova, S. K. Davidson, D. J. Faulkner, and M. G. Haygood, "Identification of the Antifungal Peptide-Containing Symbiont of the Marine Sponge

NOTES

Theonella swinhoei as a Novel δ-Proteobacterium, '*Candidatus* Entotheonella palauensis,'" *Marine Biology* 136, no. 6 (July 2000): 969–77, doi.org/10.1007/s002270000273.

15 **Like corals, some sponge species:** Yoo Kyung Lee, Jung-Hyun Lee, and Hong Kum Lee, "Microbial Symbiosis in Marine Sponges," *Journal of Microbiology* 39, no. 4 (December 2001): 254–64.

Meggie Hudspith, Laura Rix, Michelle Achlatis, Jeremy Bougoure, Paul Guagliardo, Peta L. Clode, Nicole S. Webster, Gerard Muyzer, Mathieu Pernice, and Jasper M. de Goeij, "Subcellular View of Host-Microbiome Nutrient Exchange in Sponges: Insights into the Ecological Success of an Early Metazoan-Microbe Symbiosis," *Microbiome* 9 (February 2021): 44, doi.org/10.1186/s40168-020-00984-w.

Clive R. Wilkinson and Peter Fay, "Nitrogen Fixation in Coral Reef Sponges with Symbiotic Cyanobacteria," *Nature* 279, no. 5713 (June 1979): 527–29, doi.org/10.1038/279527a0.

15 **They produce a wide range:** Selvakumar Dharmaraj, "Marine *Streptomyces* as a Novel Source of Bioactive Substances," *World Journal of Microbiology and Biotechnology* 26, no. 12 (December 2010): 2123–39, doi.org/10.1007/s11274-010-0415-6.

16 **In one study, ninety-four:** Selvakumar Dharmaraj and Alagarsamy Sumantha, "Bioactive Potential of *Streptomyces* Associated with Marine Sponges," *World Journal of Microbiology and Biotechnology* 25, no. 11 (November 2009): 1971–79, doi.org/10.1007/s11274-009-0096-1.

16 **"about 70% of the naturally":** Sheila Marie Pimentel-Elardo, Svitlana Kozytska, Tim S. Bugni, Chris M. Ireland, Heidrun Moll, and Ute Hentschel, "Anti-Parasitic Compounds from *Streptomyces sp.* Strains Isolated from Mediterranean Sponges," *Marine Drugs* 8, no. 2 (February 2010): 373–80, doi.org/10.3390/md8020373.

16 **Scientists discovered that *Streptomyces*:** Imke Schneemann, Inga Kajahn, Birgit Ohlendorf, Heidi Zinecker, Arlette Erhard, Kerstin Nagel, Jutta Wiese, and Johannes F. Imhoff, "Mayamycin, a Cytotoxic Polyketide from a *Streptomyces* Strain Isolated from the Marine Sponge *Halichondria panicea*," *Journal of Natural Products* 73, no. 7 (July 2010): 1309–12, doi.org/10.1021/np100135b.

NOTES

17 **In 1986, scientists isolated:** Yoshimasa Hirata and Daisuke Uemura, "Halichondrins—Antitumor Polyether Macrolides from a Marine Sponge," *Pure and Applied Chemistry* 58, no. 5 (1986): 701–10, doi.org/10.1351/pac198658050701.

Ruoli Bai, Kenneth D. Paull, Cherry L. Herald, Louis Malspeis, George R. Pettit, and Ernest Hamel, "Halichondrin B and Homohalichondrin B, Marine Natural Products Binding in the Vinca Domain of Tubulin. Discovery of Tubulin-Based Mechanism of Action by Analysis of Differential Cytotoxicity Data," *Journal of Biological Chemistry* 266, no. 24 (August 1991): 15882–89.

Armin Bauer, "Story of Eribulin Mesylate: Development of the Longest Drug Synthesis," in *Synthesis of Heterocycles in Contemporary Medicinal Chemistry*, ed. Zdenko Časar (Cham, Switzerland: Springer International, 2016), 209–70, doi.org/10.1007/7081_2016_201.

Thomas D. Aicher, Keith R. Buszek, Francis G. Fang, Craig J. Forsyth, Sun Ho Jung, Yoshito Kishi, Michael C. Matelich, Paul M. Scola, Denice M. Spero, and Suk Kyoon Yoon, "Total Synthesis of Halichondrin B and Norhalichondrin B," *Journal of the Amiercan Chemical Society* 114, no. 8 (April 1992): 3162–64, doi.org/10.1021/ja00034a086.

Wanjun Zheng, Boris M. Seletsky, Monica H. Palme, Paul J. Lydon, Lori A. Singer, Charles E. Chase, Charles A. Lemelin et al., "Macrocyclic Ketone Analogues of Halichondrin B," *Bioorganic & Medicinal Chemistry Letters* 14, no. 22 (November 2004): 5551–54, doi.org/10.1016/j.bmcl.2004.08.069.

Timothy K. Huyck, William Gradishar, Fil Manuguid, and Peter Kirkpatrick, "Eribulin Mesylate," *Nature Reviews Drug Discovery* 10, no. 3 (March 2011): 173–74, doi.org/10.1038/nrd3389.

18 **Dysideanone D inhibits growth:** 周同永 and 周国泰, New sesquiterpene quinine compound in Dysidea avara and application thereof, China Patent CN104478688A, filed November 24, 2014, April 1, 2015.

Pravin Shinde, Paromita Banerjee, and Anita Mandhare, "Marine Natural Products as Source of New Drugs: A Patent Review (2015–2018)," *Expert Opinion on Therapeutic Patents* 29, no. 4 (2019): 283–309, doi.org/10.1080/13543776.2019.1598972.

NOTES

18 **In my mind, working:** Angela E. Douglas, "The Molecular Basis of Bacterial–Insect Symbiosis," *Journal of Molecular Biology* 426, no. 23 (November 2014): 3830–37, doi.org/10.1016/j.jmb.2014.04.005.

19 **Many invertebrates besides sponges:** Margaret McFall-Ngai, Michael G. Hadfield, Thomas C. G. Bosch, Hannah V. Carey, Tomislav Domazet-Lošo, Angela E. Douglas, Nicole Dubilier, et al., "Animals in a Bacterial World, a New Imperative for the Life Sciences," *PNAS* 110, no. 9 (February 2013): 3229–36, doi.org/10.1073/pnas.1218525110.

M. Sofia Gil-Turnes, Mark E. Hay, and William Fenical, "Symbiotic Marine Bacteria Chemically Defend Crustacean Embryos from a Pathogenic Fungus," *Science* 246, no. 4926 (October 1989): 116–18, doi.org/10.1126/science.2781297.

Michael G. Hadfield, "Biofilms and Marine Invertebrate Larvae: What Bacteria Produce That Larvae Use to Choose Settlement Sites," *Annual Review of Marine Science* 3 (January 2011): 453–70, doi.org/10.1146/annurev-marine-120709-142753.

Irene Miguel-Aliaga, Heinrich Jasper, and Bruno Lemaitre, "Anatomy and Physiology of the Digestive Tract of *Drosophila melanogaster*," *Genetics* 210, no. 2 (October 2018): 357–96, doi.org/10.1534/genetics.118.300224.

19 **An estimated 37 percent:** McFall-Ngai et al. "Animals in a Bacterial World."

Tomislav Domazet-Lošo and Diethard Tautz, "An Ancient Evolutionary Origin of Genes Associated with Human Genetic Diseases," *Molecular Biology and Evolution* 25, no. 12 (December 2008): 2688–707, doi.org/10.1093/molbev/msn214.

21 **Because oceans represent:** Richard K. Grosberg, Geerat J. Vermeij, and Peter C. Wainwright, "Biodiversity in Water and on Land," *Current Biology* 22, no. 21 (November 2012): R900–3, doi.org/10.1016/j.cub.2012.09.050.

21 **The widely recognized value:** Caitlin Keating-Bitonti, "The Biodiversity beyond National Jurisdiction Agreement (High Seas Treaty)," IF12283, *Congressional Research Service*, July 14, 2023, sgp.fas.org/crs/misc/IF12283.pdf.

NOTES

CHAPTER 2: THE CORAL'S CASTLE

26 **Coral's symbiosis with zooxanthellae:** Mikołaj K. Zapalski and Błażej Berkowski, "The Silurian Mesophotic Coral Ecosystems: 430 Million Years of Photosymbiosis," *Coral Reefs* 38, no. 1 (February 2019): 137–47, doi.org/10.1007/s00338-018-01761-w.

28 **After returning home, Darwin:** Charles Darwin, *The Structure and Distribution of Coral Reefs* (London: Smith, Elder, 1842).

29 **On wave-swept shores:** Filippo Ferrario, Michael W. Beck, Curt D. Storlazzi, Fiorenza Micheli, Christin C. Shepard, and Laura Airoldi, "The Effectiveness of Coral Reefs for Coastal Hazard Risk Reduction and Adaptation," *Nature Communications* 5 (2014): 3794, doi.org/10.1038/ncomms4794.

29 **While we don't have data:** John A. Chamberlain Jr., "Mechanical Properties of Coral Skeleton: Compressive Strength and Its Adaptive Significance," *Paleobiology* 4, no. 4 (Fall 1978): 419–35, doi.org/10.1017/S0094837300006163.

29 **Corals don't simply patch:** Gabriela A. Farfan, Amy Apprill, Anne Cohen, Thomas M. DeCarlo, Jeffrey E. Post, Rhian G. Waller, and Colleen M. Hansel, "Crystallographic and Chemical Signatures in Coral Skeletal Aragonite," *Coral Reefs* 41, no. 1 (February 2022): 19–34, doi.org/10.1007/s00338-021-02198-4.

30 **We know sunlight has:** Thomas F. Goreau and Nora I. Goreau, "The Physiology of Skeleton Formation in Corals. II. Calcium Deposition by Hermatypic Corals under Various Conditions in the Reef," *Biological Bulletin* 117, no. 2 (October 1959): 239–50.

31 **Scientists recently discovered that:** Susana Enríquez, Eugenio R. Méndez, and Roberto Iglesias-Prieto, "Multiple Scattering on Coral Skeletons Enhances Light Absorption by Symbiotic Algae," *Limnology and Oceanography* 50, no. 4 (July 2005): 1025–32, doi.org/10.4319/lo.2005.50.4.1025.

32 **They designed 3D printed:** Daniel Wangpraseurt, Shangting You, Farooq Azam, Gianni Jacucci, Olga Gaidarenko, Mark Hildebrand, Michael Kühl, et al., "Bionic 3D Printed Corals," *Nature Communications* 11 (2020): 1748, doi.org/10.1038/s41467-020-15486-4.

NOTES

32 **Researchers have discovered that coral:** H. Ehrlich, P. Etnoyer, S. D. Litvinov, M. M. Olennikova, H. Domaschke, T. Hanke, R. Born, H. Meissner, and H. Worch, "Biomaterial Structure in Deep-Sea Bamboo Coral (Anthazoa: Gorgonacea: Isididae): Perspectives for the Development of Bone Implants and Templates for Tissue Engineering," in "Biomaterials," special issue, *Materialwissenschaft und Werkstofftechnik* 37, no. 6 (June 2006): 552–57, doi.org/10.1002/mawe.200600036.

E. Damien and P. A. Revell, "Coralline Hydroxyapatite Bone Graft Substitute: A Review of Experimental Studies and Biomedical Applications," *Journal of Applied Biomaterials & Biomechanics* 2, no. 2 (May–August 2004): 65–73, https://journals.sagepub.com/doi/abs/10.1177/228080000400200201.

M. Sivakumar, T. S. Sampath Kumar, K. L. Shantha, and K. Panduranga Rao, "Development of Hydroxyapatite Derived from Indian Coral," *Biomaterials* 17, no. 17 (1996): 1709–14, doi.org/10.1016/0142-9612(96)87651-4.

33 **These composite skeletons can provide:** H. Ehrlich et al., "Biomaterial Structure in Deep-Sea Bamboo Coral."

33 **The common ancestor:** Catherine S. McFadden, Andrea M. Quattrini, Mercer R. Brugler, Peter F. Cowman, Luisa F. Dueñas, Marcelo V. Kitahara, David A. Paz-García, James D. Reimer, and Estefanía Rodríguez, "Phylogenomics, Origin, and Diversification of Anthozoans (Phylum Cnidaria)," *Systematic Biology* 70, no. 4 (July 2021): 635–47, doi.org/10.1093/sysbio/syaa103.

33 **Colonial growth forms:** Artem Kouchinsky, Stefan Bengtson, Brue Runnegar, Christian Skovsted, Michael Steiner, and Michael Vendrasco, "Chronology of Early Cambrian Biomineralization," *Geological Magazine* 149, no. 2 (March 2012): 221–51, doi.org/10.1017/S0016756811000720.

33 **In 2022, a newly:** F. S. Dunn, C. G. Kenchington, L. A. Parry, J. W. Clark, R. S. Kendall, and P. R. Wilby, "A Crown-Group Cnidarian from the Ediacaran of Charnwood Forest, UK," *Nature Ecology & Evolution* 6 (2022): 1095–104, doi.org/10.1038/s41559-022-01807-x.

34 **The large and widespread:** McFadden et al., "Phylogenomics, Origin, and Diversification of Anthozoans."

NOTES

36 **It was known that zooxanthellae:** Leslie Roberts, "Coral Bleaching Threatens Atlantic Reefs," *Science* 238, no. 4831 (November 1987): 1228–29, doi.org/10.1126/science.238.4831.1228.

37 **By the time El Niño:** Clive R. Wilkinson and G. Hodgson, "Coral Reefs and the 1997–1998 Mass Bleaching and Mortality," *Nature & Resources* 35, no. 2 (April–June 1999): 16–25.

37 **Those of us who studied:** O. Hoegh-Guldberg, P. J. Mumby, A. J. Hooten, R. S. Steneck, P. Greenfield, E. Gomez, C. D. Harvell, et al., "Coral Reefs under Rapid Climate Change and Ocean Acidification," *Science* 318, no. 5857 (December 2007): 1737–42, doi.org/10.1126/science.1152509.

C. Drew Harvell, Charles E. Mitchell, Jessica R. Ward, Sonia Altizer, Andrew P. Dobson, Richard S. Ostfeld, and Michael D. Samuel, "Climate Warming and Disease Risks for Terrestrial and Marine Biota," *Science* 296, no. 5576 (June 2002): 2158–62, doi.org/10.1126/science.1063699.

C. D. Harvell, K. Kim, J. M. Burkholder, R. R. Colwell, P. R. Epstein, D. J. Grimes, E. E. Hofmann, et al., "Emerging Marine Diseases—Climate Links and Anthropogenic Factors," *Science* 285, no. 5433 (September 1999): 1505–10, doi.org/10.1126/science.285.5433.1505.

41 **My postdoctoral scholar:** Joleah B. Lamb, David H. Williamson, Garry R. Russ, and Bette L. Willis, "Protected Areas Mitigate Diseases of Reef-Building Corals by Reducing Damage from Fishing," *Ecology* 96, no. 9 (September 2015): 2555–67, doi.org/10.1890/14-1952.1.

Joleah B. Lamb, Bette L. Willis, Evan A. Fiorenza, Courtney S. Couch, Robert Howard, Douglas N. Rader, James D. True, et al., "Plastic Waste Associated with Disease on Coral Reefs," *Science* 359, no. 6374 (January 2018): 460–62, doi.org/10.1126/science.aar3320.

46 **The results were clear:** Joleah B. Lamb, Jeroen A. J. M. van de Water, David G. Bourne, Craig Altier, Margaux Y. Hein, Evan A. Fiorenza, Nur Abu, Jamaluddin Jompa, and C. Drew Harvell, "Seagrass Ecosystems Reduce Exposure to Bacterial Pathogens of Humans, Fishes, and Invertebrates," *Science* 355, no. 6326 (February 2017): 731–33, doi.org/10.1126/science.aal1956.

NOTES

46 **We went on to record:** Reyn M. Yoshioka, Catherine J. S. Kim, Allison M. Tracy, Rebecca Most, and C. Drew Harvell, "Linking Sewage Pollution and Water Quality to Spatial Patterns of *Porites lobata* Growth Anomalies in Puako, Hawaii," *Marine Pollution Bulletin* 104, no. 1–2 (March 2016): 313–21, doi.org/10.1016/j.marpolbul.2016.01.002.

46 **With the help of Seattle's:** Phoebe Dawkins, Evan A. Fiorenza, Jeffrey Gaeckle, Jennifer Lanksbury, Jeroen van de Water, William Feeney, C. Drew Harvell, and Joleah B. Lamb. "Global Seagrass Ecosystems as Green Urban Infrastructure to Mediate Human Pathogens in Food from the Sea," *Nature Sustainability* (August 2024).

46 **Seagrass meadows:** Lynda V. Mapes, "Researchers Discover Eelgrass Superpower in Puget Sound," *The Seattle Times*, August 14, 2024, https://www.seattletimes.com/seattle-news/climate-lab/researchers-discover-eelgrass-superpower-in-puget-sound.

47 **In 2008, following multiple:** Kent E. Carpenter, Muhammad Abrar, Greta Aeby, Richard B. Aronson, Stuart Banks, Andrew Bruckner, Angel Chiriboga, et al., "One-Third of Reef-Building Corals Face Elevated Extinction Risk from Climate Change and Local Impacts," *Science* 321, no. 5888 (July 2008): 560–63, doi.org/10.1126/science.1159196.

47 **A study in 2021:** Terry P. Hughes, James T. Kerry, Andrew H. Baird, Sean R. Connolly, Andreas Dietzel, C. Mark Eakin, Scott F. Heron, et al., "Global Warming Transforms Coral Reef Assemblages," *Nature* 556 (April 2018): 492–96, doi.org/10.1038/s41586-018-0041-2.

Terry P. Hughes, James T. Kerry, Sean R. Connolly, Jorge G. Álvarez-Romero, C. Mark Eakin, Scott F. Heron, Migdonio A. Gonzalez, and Joanne Moneghetti, "Emergent Properties in the Responses of Tropical Corals to Recurrent Climate Extremes," *Current Biology* 31, no. 23 (December 2021): P5393–99.E3, doi.org/10.1016/j.cub.2021.10.046.

49 **Scientists and engineers are looking:** Ove Hoegh-Guldberg, Eliza Northrop, and Jane Lubchenco, "The Ocean Is Key to Achieving Climate and Societal Goals," *Science* 365, no. 6460 (September 2019): 1372–74, doi.org/10.1126/science.aaz4390.

NOTES

CHAPTER 3: THE SEA FAN'S ANCIENT DEFENSES

59 **Comparing the before-and-after time:** Thomas H. Suchanek, Robert C. Carpenter, Jon D. Witman, and C. Drew Harvell, "Sponges as Important Space Competitors in Deep Coral Reef Communities," in *The Ecology of Deep and Shallow Coral Reefs*, ed. Marjorie L. Reaka-Kudla (Rockville, MD: US Department of Commerce, National Oceanic and Atmospheric Administration, Oceanic and Atmospheric Research, Office of Undersea Research, 1983), 55–61.

61 **The changing climate:** C. Drew Harvell, Charles E. Mitchell, Jessica R. Ward, Sonia Altizer, Andrew P. Dobson, Richard S. Ostfeld, and Michael D. Samuel, "Climate Warming and Disease Risks for Terrestrial and Marine Biota," *Science* 296, no. 5576 (June 2022): 2158–61, doi.org/10.1126/science.1063699.

62 **Octocorals, along with sponges:** Inês Raimundo, Sandra G. Silva, Rodrigo Costa, and Tina Keller-Costa, "Bioactive Secondary Metabolites from Octocoral-Associated Microbes—New Chances for Blue Growth," in "Bioactive Compounds from Marine Microbes—II," special issue, *Marine Drugs* 16, no. 12 (December 2018): 485, doi.org/10.3390/md1 6120485.

Leo C. Vining, "Functions of Secondary Metabolites," *Annual Review of Microbiology* 44 (October 1990): 395–427, doi.org/10.1146/annurev.mi.44.100190.002143.

Paul W. Sammarco and John C. Coll, "Chemical Adaptations in the Octocorallia: Evolutionary Considerations," *Marine Ecology Progress Series* 88 (1992): 93–104, doi.org/10.3354/meps088093.

63 **Five years after our research:** Jaime Garzón-Ferreira and Sven Zea, "A Mass Mortality of *Gorgonia ventalina* (Cnidaria: Gorgoniidae) in the Santa Marta Area, Caribbean Coast of Colombia," *Bulletin of Marine Science* 50, no. 3 (May 1992): 522–26.

I. Nagelkerken, K. Buchan, G. W. Smith, K. Bonair, P. Bush, J. Garzón-Ferreira, L. Botero, et al., "Widespread Disease in Caribbean Sea Fans: II. Patterns of Infection and Tissue Loss," *Marine Ecology Progress Series* 160 (1997): 255–63, doi.org/10.3354/meps 160255.

NOTES

66 **In the summer of 1996:** Garriet W. Smith, Lisa D. Ives, Ivan A. Nagelkerken, and Kim B. Ritchie, "Caribbean Sea-Fan Mortalities," *Nature* 383 (1996): 487, doi.org/10.1038/383487a0.

68 **Under a microscope, we could:** L. E. Petes, C. Drew Harvell, E. C. Peters, M. A. H. Webb, and K. M. Mullen, "Pathogens Compromise Reproduction and Induce Melanization in Caribbean Sea Fans," *Marine Ecology Progress Series* 264 (December 2003): 167–71, doi.org/10.3354/meps264167.

Laura D. Mydlarz, Sally F. Holthouse, Esther C. Peters, and C. Drew Harvell, "Cellular Responses in Sea Fan Corals: Granular Amoebocytes React to Pathogen and Climate Stressors," *PLoS One* 3, no. 3 (March 2008): e1811, doi.org/10.1371/journal.pone.0001811.

69 **Prophenoloxidase and the cascade:** Laura D. Mydlarz, Laura E. Jones, and C. Drew Harvell, "Innate Immunity, Environmental Drivers, and Disease Ecology of Marine and Freshwater Invertebrates," *Annual Review of Ecology, Evolution, and Systematics* 37 (December 2006): 251–88, doi.org/10.1146/annurev.ecolsys.37.091305.110103.

69 **Laura had just added corals:** Laura D. Mydlarz and C. Drew Harvell, "Peroxidase Activity and Inducibility in the Sea Fan Coral Exposed to a Fungal Pathogen," *Comparative Biochemistry and Physiology Part A: Molecular & Integrative Physiology* 146, no. 1 (January 2007): 54–62, doi.org/10.1016/j.cbpa.2006.09.005.

70 **Although we detected antimicrobial:** P. R. Jensen, C. D. Harvell, K. Wirtz, and W. Fenical, "Antimicrobial Activity of Extracts of Caribbean Gorgonian Corals," *Marine Biology* 125, no. 2 (April 1996): 411–19, doi.org/10.1007/BF00346321.

71 **This pattern of survival:** Kiho Kim and C. Drew Harvell, "The Rise and Fall of a Six-Year Coral-Fungal Epizootic," in "Ecology and Evolution of Host-Pathogen Interactions in Natural Populations: A Symposium Organized by Drew Harvell," *American Naturalist* 164, no. S5 (November 2004): S52–63, doi.org/10.1086/424609.

John F. Bruno, Stephen P. Ellner, Ivana Vu, Kiho Kim, and C. Drew Harvell, "Impacts of Aspergillosis on Sea Fan Coral Demography: Modeling a Moving Target," *Ecological Monographs* 81, no. 1 (February 2011): 123–39, doi.org/10.1890/09-1178.1.

72 **The properties of our innate:** Jakob Suckale, Robert B. Sim, and Alister W. Dodds, "Evolution of Innate Immune Systems," *Biochemistry and Molecular Biology Education* 33, no. 3 (May 2005): 177–83, doi.org/10.1002/bmb.2005.494033032466.

C. Toledo-Hernández and C. P. Ruiz-Diaz, "The Immune Responses of the Coral," *Invertebrate Survival Journal* 11, no. 1 (2014): 319–28.

Douglas R. McDonald and Ofer Levy, "Innate Immunity," in *Clinical Immunology: Principles and Practice*, ed. Robert R. Rich, Thomas A. Fleisher, William T. Shearer, Harry W. Schroeder Jr., Anthony J. Frew, and Cornelia M. Weyland, 5th ed. (London: Elsevier, 2019), 39–53.e1, doi.org/10.1016/B978-0-7020-6896-6.00003-X.

73 **One interesting twist:** Domenico Birra, Maurizio Benucci, Luigi Landolfi, Anna Merchionda, Gabriella Loi, Patrizia Amato, Gaetano Licata, Luca Quartuccio, Massimo Triggiani, and Paolo Moscato, "COVID 19: A Clue from Innate Immunity," *Immunologic Research* 68, no. 3 (June 2020): 161–68, doi.org/10.1007/s12026-020-09137-5.

73 **Because innate immunity wanes:** Giorgio Costagliola, Erika Spada, and Rita Consolini, "Age-Related Differences in the Immune Response Could Contribute to Determine the Spectrum of Severity of COVID-19," *Immunity, Inflammation and Disease* 9, no. 2 (June 2021): 331–39, doi.org/10.1002/iid3.404.

73 **In an article entitled:** Konstantin Chumakov, Michael S. Avidan, Christine S. Benn, Stefano M. Bertozzi, Lawrence Blatt, Angela Y. Chang, Dean T. Jamison, et al., "Old Vaccines for New Infections: Exploiting Innate Immunity to Control COVID-19 and Prevent Future Pandemics," *PNAS* 118, no. 21 (May 2021): e2101718118, doi.org/10.1073/pnas.2101718118.

73 **One example they cite:** Johanneke Kleinnijenhuis, Jessica Quintin, Frank Preijers, Leo A. B. Joosten, Daniela C. Ifrim, Sadia Saeed, Cor Jacobs, et al., "Bacille Calmette-Guérin Induces NOD2-Dependent Nonspecific Protection from Reinfection via Epigenetic Reprogramming of Monocytes" *PNAS* 109, no. 43 (September 2012): 17537–42, doi.org/10.1073/pnas.1202870109.

74 **Early in the COVID-19 outbreak:** Chumakov et al., "Old Vaccines for New Infections."

NOTES

CHAPTER 4: THE SEA SLUG'S STING

79 **They fire in less:** T. Holstein and P. Tardent, "An Ultrahigh-Speed Analysis of Exocytosis: Nematocyst Discharge," *Science* 233, no. 4638 (February 1982): 830–33, doi.org/10.1126/science.6695186.

79 **The harpoons launch:** Timm Nüchter, Martin Benoit, Ulrike Engel, Suat Özbek, and Thomas W. Holstein, "Nanosecond-Scale Kinetics of Nematocyst Discharge," *Current Biology* 16, no. 9 (May 2006): R316–18, doi.org/10.1016/j.cub.2006.03.089.

83 **The color palette:** Marta Pola, Yara Tibiriçá, and Juan Lucas Cervera, "Psychedelics Sea Slugs: Observations on Colour Ontogeny in Two Nudibranch Species from the Genus Nembrotha (Doridina: Polyceridae)," *Scientia Marina* 87, no. 3 (September 2023): e072, doi.org/10.3989/scimar.05371.072.

87 **I developed the hypothesis:** C. Drew Harvell and Thomas H. Suchanek, "Partial Predation on Tropical Gorgonians by *Cyphoma gibbosum* (Gastropoda)," *Marine Ecology Progress Series* 38 (May 1987): 37–44.

87 **We went on to learn:** Gordon W. Gribble, "Newly Discovered Naturally Occurring Organohalogens," *Arkivoc, pt. i* (2018): 372–410, doi.org/10.24820/ark.5550190.p010.610.

C. Drew Harvell, William Fenical, Vassilios Roussis, Jennifer L. Ruesink, Catherine C. Griggs, and Charles H. Greene, "Local and Geographic Variation in the Defensive Chemistry of a West Indian Gorgonian Coral (*Briareum asbestinum*)," *Marine Ecology Progress Series* 93, no. 1/2 (February 1993): 165–73.

87 **Among the first scientists:** Thomas Hartmann, "The Lost Origin of Chemical Ecology in the Late 19th Century," *PNAS* 105, no. 12 (March 2008): 4541–46, doi.org/10.1073/pnas.0709231105.

88 **As a growing caterpillar:** Anurag Agrawal, *Monarchs and Milkweed* (Princeton, NJ: Princeton University Press, 2017), doi.org/10.1515/9781400884766.

88 **hypothesis of toxin dilution:** Richard Karban, Claire Karban, Mikaela Huntzinger, Ian Pearse, and Gregory Crutsinger, "Diet Mixing Enhances the Performance of a Generalist Caterpillar, *Platyprepia virginalis*,"

NOTES

Ecological Entomology 35, no. 1 (January 2010): 92–99, doi.org/10.1111 /j.1365-2311.2009.01162.x.

90 **In 1843, the famous:** Joshua Alder and Albany Hancock, *A Monograph of the British Nudibranchiate Mollusca: With Figures of All the Species* (London: Ray Society, 1845–1855), doi.org/10.5962/bhl.title.65015.

90 **He surmised that the tiny:** T. Strethill Wright, "On the Urticating Filaments of the Eolidae," *Quarterly Journal of Microscopical Science* S2-3, no. 9 (January 1863): 52–55, doi.org/10.1242/jcs.s2-3.9.52.

90 **Everyone agreed the threads:** R. Bergh, "On the Existence of the Urticating Filaments in the Mollusca," *Quarterly Journal of Microscopical Science* S2-2, no. 8 (October 1862): 274–77, doi.org/10.1242/jcs.s2 -2.8.274.

90 **Two other prominent biologists:** Otto C. Glaser, *The Nematocysts of Eolids* (Baltimore: Williams & Wilkins, 1910).

91 **Finally, in 1903, Oxford lecturer:** G. H. Grosvenor, "On the Nematocysts of Aeolids," *Proceedings of the Royal Society of London* 72 (January 1904): 462–86, doi.org/10.1098/rspl.1903.0075.

92 **Dr. Otto Glaser, University of Michigan:** Glaser, *The Nematocysts of Eolids*.

92 **He proposed that nudibranchs:** A. Naville, "Notes sur les Eolidiens. Un Eolidien d'eau saumâtre. Origine des nématocystes, zooxanthelles et homochromie," *Revue suisse de zoologie* 33 (1926): 251–89, doi.org /10.5962/bhl.part.117605.

92 **Then, in 1984, Paul Greenwood:** Paul G. Greenwood and Richard N. Mariscal, "Immature Nematocyst Incorporation by the Aeolid Nudibranch *Spurilla neapolitana*," *Marine Biology* 80, no. 1 (May 1984): 35–38, doi.org/10.1007/BF00393124.

Paul G. Greenwood and Richard N. Mariscal, "The Utilization of Cnidarian Nematocysts by Aeolid Nudibranchs: Nematocyst Maintenance and Release in *spurilla*," *Tissue and Cell* 16, no. 5 (1984): 719–30, doi.org/10.1016/0040-8166(84)90005-3.

92 **More recently, new studies:** R. Martin, "Management of Nematocysts in the Alimentary Tract and in Cnidosacs of the Aeolid Nudibranch

Gastropod *Cratena peregrina*," *Marine Biology* 143, no. 3 (September 2003): 533–41, doi.org/10.1007/s00227-003-1078-8.

Paul G. Greenwood, "Acquisition and Use of Nematocysts by Cnidarian Predators," *Toxicon* 54, no. 8 (December 2009): 1065–70, doi.org/10.1016/j.toxicon.2009.02.029.

Jessica A. Goodheart and Alexandra E. Bely, "Sequestration of Nematocysts by Divergent Cnidarian Predators: Mechanism, Function, and Evolution," *Invertebrate Biology* 136, no. 1 (March 2017): 75–91, doi.org/10.1111/ivb.12154.

92 **To this day, the details:** Jessica A. Goodheart, Vanessa Barone, and Deirdre C. Lyons, "Movement and Storage of Nematocysts across Development in the Nudibranch Berghia stephanieae (Valdés, 2005)," *Frontiers in Zoology* 19 (April 2022): 16, doi.org/10.1186/s12983-022-00460-1.

93 **Ctenophores, flatworms, acoel flatworms:** Goodheart and Bely, "Sequestration of Nematocysts by Divergent Cnidarian Predators."

93 **Within the nudibranchs, it happens:** Goodheart and Bely, "Sequestration of Nematocysts by Divergent Cnidarian Predators."

93 **The herbivorous sea slugs:** Vesa Havurinne and Esa Tyystjärvi, "Photosynthetic Sea Slugs Induce Protective Changes to the Light Reactions of the Chloroplasts They Steal from Algae," *eLife* 9 (2020): e57389, doi.org/10.7554/eLife.57389.

94 *Berghia* **have evolved:** Goodheart, Barone, and Lyons, "Movement and Storage of Nematocysts."

95 **The blue dragon is a:** C. Drew Harvell, "Voyage of the Blue Dragon," Drew Harvell, March 25, 2019, drewharvell.com/2019/03/25/voyage-of-the-blue-dragon.

96 **One recent study shows:** Dana Obermann, Ulf Bickmeyer, and Heike Wägele, "Incorporated Nematocysts in *Aeolidiella stephanieae* (Gastropoda, Opisthobranchia, Aeolidoidea) Mature by Acidification Shown by the pH Sensitive Fluorescing Alkaloid Ageladine A," *Toxicon* 60, no. 6 (November 2012): 1108–16, doi.org/10.1016/j.toxicon.2012.08.003.

NOTES

97 **Pigs are a species:** Megan Sykes, "Developing Pig-to-Human Organ Transplants," *Science* 378, no. 6616 (October 2022): 135–36, doi.org/10.1126/science.abo7935.

97 **Researchers have successfully transplanted:** Vladimir A. Morozov, Shaun Wynyard, Shinichi Matsumoto, Adrian Abalovich, Joachim Denner, and Robert Eillott, "No PERV Transmission during a Clinical Trial of Pig Islet Cell Transplantation," *Virus Research* 227 (January 2017): 34–40, doi.org/10.1016/j.virusres.2016.08.012.

97 **Active research is focusing:** Paige M. Porrett, Babak J. Orandi, Vineeta Kumar, Julie Houp, Douglas Anderson, A. Cozette Killian, Vera Hauptfeld-Dolejsek, et al., "First Clinical-Grade Porcine Kidney Xenotransplant Using a Human Decedent Model," *American Journal of Transplantation* 22, no. 4 (April 2022): 1037–53, doi.org/10.1111/ajt.16930.

CHAPTER 5: THE GIANT CLAM'S LIGHT TRICK

102 **So our work was:** Melissa Garren, Laurie Raymundo, James Guest, C. Drew Harvell, and Farooq Azam, "Resilience of Coral-Associated Bacterial Communities Exposed to Fish Farm Effluent," *PLoS ONE* 4, no. 10 (October 2009): e7319, doi.org/10.1371/journal.pone.0007319.

105 **This was *Tridacna gigas*:** "Giant Clam (Tridacna spp.)," NOAA Fisheries, last modified September 15, 2022, fisheries.noaa.gov/species/giant-clam-tridacna-spp.

106 **The key to their large size:** C. M. Yonge, "Giant Clams," *Scientific American* 232, no. 4 (April 1975): 96–105.

106 **These clams have other tricks:** Pamela Soo and Peter A. Todd, "The Behaviour of Giant Claims (Bivalvia: Cardiidae: Tridacninae)," *Marine Biology* 161, no. 12 (December 2014): 2699–717, doi.org/10.1007/s00227-014-2545-0.

Richard D. Braley, "Reproduction in the Giant Clams *Tridacna gigas* and *T. derasa* In Situ on the North-Central Great Barrier Reef, Australia, and Papua New Guinea," *Coral Reefs* 3, no. 4 (December 1984): 221–27, doi.org/10.1007/BF00288258.

NOTES

107 **One threat is a cadre:** H. Govan, L. Y. Fabro, and E. Ropeti, "Controlling Predators of Cultured Tridacnid Clams," *Proceedings-Australian Centre for International Agricultural Research* (1993).

F. E. Perron, G. A. Heslinga, and J. O. Fagolimul, "The Gastropod *Cymatium muricinum*, a Predator on Juvenile Tridacnid Clams," *Aquaculture* 48, no. 3–4 (September 1985): 211–21, doi.org/10.1016/0044-8486(85)90125-5.

108 **Its reliance on energy:** D. W. Klumpp and C. L. Griffiths, "Contributions of Phototrophic and Heterotrophic Nutrition to the Metabolic and Growth Requirements of Four Species of Giant Clam (Tridacnidae)," *Marine Ecology Progress Series* 115 (December 1994): 103–15.

109 **While corals also have evolved:** D. J. Griffiths, H. Winsor, and T. Luongvan, "Iridophores in the Mantle of Giant Clams," *Australian Journal of Zoology* 40, no. 3 (1992): 319–26, doi.org/10.1071/ZO9920319.

110 **The iridocytes allow clams:** Amanda L. Holt, Sanaz Vahidinia, Yakir Luc Gagnon, Daniel E. Morse, and Alison M. Sweeney, "Photosymbiotic Giant Clams Are Transformers of Solar Flux," *Journal of the Royal Society Interface* 11, no. 101 (December 2014): 20140678, doi.org/10.1098/rsif.2014.0678.

Amitabh Ghoshal, Elizabeth Eck, Michael Gordon, and Daniel E. Morse, "Wavelength-Specific Forward Scattering of Light by Bragg-Reflective Iridocytes in Giant Clams," *Journal of the Royal Society Interface* 13, no. 120 (July 2016): 20160285, doi.org/10.1098/rsif.2016.0285.

110 **These cells are arranged in:** Holt et al., "Photosymbiotic Giant Clams Are Transformers of Solar Flux."

110 **Interestingly, the columns containing:** John H. Norton, Malcom A. Shepherd, Helen M. Long, and William K. Fitt, "The Zooxanthellal Tubular System in the Giant Clam," *Biological Bulletin* 183, no. 3 (December 1992): 503–6, doi.org/10.2307/1542028.

111 **The key superpower that is:** Susann Rossbach, Ram Chandra Subedi, Tien Khee Ng, Boon S. Ooi, and Carlos M. Duarte, "Iridocytes Mediate Photonic Cooperation between Giant Clams (Tridacninae) and Their Photosynthetic Symbionts," *Frontiers in Marine Science* 7 (June 2020): 465, doi.org/10.3389/fmars.2020.00465.

NOTES

Susann Rossbach, Sebastian Overmans, Altynay Kaidarova, Jürgen Kosel, Susana Agustí, and Carlos M. Duarte, "Giant Clams in Shallow Reefs: UV-Resistance Mechanisms of Tridacninae in the Red Sea," *Coral Reefs* 39, no. 5 (October 2020): 1345–60, doi.org/10.1007/s00338-020-01968-w.

Amanda L. Holt, Lincoln F. Rehm, and Alison M. Sweeney, "The Giant Clam Photosymbiosis Is a Physically Optimal Photoconversion System for the Most Intense Sunlight on Earth," preprint, posted March 1, 2023, doi.org/10.1101/2023.02.28.530416.

111 **"the guanine palettes not only":** Susann Rossbach, "A Different Slant of Light," *KAUST Discovery*, July 5, 2020, discovery.kaust.edu.sa/en/article/6366/a-different-slant-of-light/.

112 **Baby fringe-fingered lizards:** Gan Zhang, Venkata Jayasurya Yallapragada, Topaz Halperin, Avital Wagner, Michal Shemesh, Alexander Upcher, Iddo Pinkas, Harry L. O. McClelland, Dror Hawlena, and Benjamin A. Palmer, "Lizards Exploit the Changing Optics of Developing Chromatophore Cells to Switch Defensive Colors during Ontogeny," *PNAS* 120, no. 18 (April 2023): e2215193120, doi.org/10.1073/pnas.2215193120.

112 **Glass frogs reflect light:** Carlos Taboada, Jesse Delia, Maomao Chen, Chenshuo Ma, Xiaorui Peng, Xiaoyi Zhui, Laiming Jiang, et al., "Glassfrogs Conceal Blood in Their Liver to Maintain Transparency," *Science* 378, no. 6626 (December 2022): 1315–20, doi.org/10.1126/science.abl6620.

112 **The versatility and complexity of guanine:** Dvir Gur, Benjamin A. Palmer, Steve Weiner, and Lia Addadi, "Light Manipulation by Guanine Crystals in Organisms: Biogenic Scatterers, Mirrors, Multilayer Reflectors and Photonic Crystals," *Advanced Functional Materials* 27, no. 6 (February 2017): 1603514, doi.org/10.1002/adfm.201603514.

112 **"Evolution is so much more clever":** Cynthia Barnett, "Mysterious Glowing Clams Could Help Save the Planet," *The Atlantic*, September 10, 2018, theatlantic.com/science/archive/2018/09/the-glowing-secrets-of-giant-clams/568774.

113 **"By that I mean giant":** Jim Shelton, "Giant Clams, Pollen, and Squid Eyes—Blueprints for a Better World," *YaleNews*, April 26, 2021, news

.yale.edu/2021/04/26/giant-clams-pollen-and-squid-eyes-blueprints
-better-world.

113 **The efficiency of the cells:** Holt et al., "Photosymbiotic Giant Clams Are Transformers of Solar Flux."

Holt, Rehm, and Sweeney, "The Giant Clam Photosymbiosis."

114 **Recent studies show that where:** Rossbach et al., "Iridocytes Mediate Photonic Cooperation."

115 **Although giant clams are huge:** Ahmad Syukri bin Othman, Gideon H. S. Goh, and Peter A. Todd, "The Distribution and Status of Giant Clams (Family Tridacnidae)—a Short Review," *Raffles Bulletin of Zoology* 58, no. 1 (February 2010): 103–11.

Sue M. Wells, *Giant Clams: Status, Trade and Mariculture, and the Role of CITES in Management* (Gland, Switzerland: IUCN, 1997).

116 **When warmed, the symbionts:** Deepak Apte, Sumantha Narayana, and Sutirtha Dutta, "Impact of Sea Surface Temperature Anomalies on Giant Clam Population Dynamics in Lakshadweep Reefs: Inferences from a Fourteen Years Study," *Ecological Indicators* 107 (December 2019): 105604, doi.org/10.1016/j.ecolind.2019.105604.

118 **On the Great Barrier Reef:** Mei Lin Neo, William Eckman, Kareen Vicentuan, Serena L.-M. Teo, and Peter A. Todd, "The Ecological Significance of Giant Clams in Coral Reef Ecosystems," *Biological Conservation* 181 (January 2015): 111–23, doi.org/10.1016/j.biocon.2014.11.004.

118 **The essential new insights contributed:** Hye-Na Kim, Sanaz Vahidinia, Amanda L. Holt, Alison M. Sweeney, and Shu Yang, "Geometric Design of Scalable Forward Scatterers for Optimally Efficient Solar Transformers," *Advanced Materials* 29, no. 44 (November 2017): 1702922. doi.org/10.1002/adma.201702922.

CHAPTER 6: THE OCTOPUS'S SHAPE SHIFT

123 **The nervous system control:** Roger T. Hanlon, John W. Forsythe, and David E. Joneschild, "Crypsis, Conspicuousness, Mimicry and

NOTES

Polyphenism as Antipredator Defences of Foraging Octopuses on Indo-Pacific Coral Reefs with a Method of Quantifying Crypsis from Video Tapes," *Biological Journal of the Linnean Society* 66, no. 1 (January 1999): 1–22, doi.org/10.1111/j.1095-8312.1999.tb01914.x.

123 **The complete visible change:** Roger T. Hanlon and Chuan-Chin Chiao, "Rapid Neural Polyphenism in Cephalopods: Current Understandings and Future Challenges," in *The Oxford Handbook of Invertebrate Neurobiology,* ed. John H. Byrne (Oxford: Oxford University Press, 2018), 701–16, doi.org/10.1093/oxfordhb/9780190456757.013.30.

123 **The beginnings of the mollusks:** Winston F. Ponder and David R. Lindberg, "Molluscan Evolution and Phylogeny: An Introduction," in *Phylogeny and Evolution of the Mollusca*, ed. Winston F. Ponder and David R. Lindberg (Berkeley: University of California Press, 2008), 1–17.

123 **These first cephalopods:** Björn Kröger, Jakob Vinther, and Dirk Fuchs, "Cephalopod Origin and Evolution: A Congruent Picture Emerging from Fossils, Development and Molecules," *BioEssays* 33, no. 8 (August 2011): 602–13, doi.org/10.1002/bies.201100001.

124 **Dr. Sydney Brenner:** Robert Sanders, "Octopus Genome Reveals Cephalopod Secrets," *Berkeley News*, August 12, 2015, news.berkeley.edu/2015/08/12/octopus-genome-reveals-cephalopod-secrets.

124 **compound, vision-forming eyes:** M. J. Wells, "Nervous Control of the Heartbeat in Octopus," *Journal of Experimental Biology* 85, no. 1 (April 1980): 111–28, doi.org/10.1242/jeb.85.1.111.

Frederike D. Hanke and Almut Kelber, "The Eye of the Common Octopus (*Octopus vulgaris*)," *Frontiers in Physiology* 10 (January 2020): 1637, doi.org/10.3389/fphys.2019.01637.

124 **Technically they have nine brains:** Danna Staaf, *The Lives of Octopuses & Their Relatives* (Princeton, NJ: Princeton University Press, 2023).

124 **All three components of change:** Tessa G. Montague, "Neural Control of Cephalopod Camouflage," *Current Biology* 33, no. 20 (October 2023): R1095–100, doi.org/10.1016/j.cub.2023.08.095.

125 **Skin color is changed:** George R. R. Bell, Alan M. Kuzirian, Stephen L. Senft, Lydia M. Mäthger, Trevor J. Wardill, and Roger T. Hanlon,

"Chromatophore Radial Muscle Fibers Anchor in Flexible Squid Skin," *Invertebrate Biology* 132, no, 2 (June 2013): 120–32, doi.org/10.1111/ivb.12016.

125 **When activated, radial muscles:** Ernst Florey, "Ultrastructure and Function of Cephalopod Chromatophores," *American Zoologist* 9, no. 2 (May 1969): 429–42, doi.org/10.1093/icb/9.2.429.

125 **The activity of the chromatophores:** J. B. Messenger, "Cephalopod Chromatophores: Neurobiology and Natural History," *Biological Reviews* 76, no. 4 (November 2001): 473–528, doi.org/10.1017/S1464793 101005772.

125 **In *Octopus vulgaris*:** J. Z. Young, "The Number and Size of Nerve Cells in *Octopus*," *Proceedings of the Zoological Society of London* 140, no. 2 (March 1963): 229–54, doi.org/10.1111/j.1469-7998.1963.tb01862.x.

125 **The extra-sparkly bling:** Lydia M. Mäthger, Eric J. Denton, N. Justin Marshall, and Roger T. Hanlon, "Mechanisms and Behavioural Functions of Structural Coloration in Cephalopods," *Journal of the Royal Society Interface* 6, no. suppl_2 (April 2009): S149–63, doi.org/10.1098/rsif.2008.0366.focus.

125 **When startled, the blue-ringed octopus:** Lydia M. Mäthger, George R. R. Bell, Alan M. Kuzirian, Justine J. Allen, and Roger T. Hanlon, "How Does the Blue-Ringed Octopus (*Hapalochlaena lunulate*) Flash Its Blue Rings?," *Journal of Experimental Biology* 215, no. 21 (November 2012): 3752–57, doi.org/10.1242/jeb.076869.

125 **Specialized cells called leucophores:** Mäthger et al., "Mechanisms and Behavioural Functions of Structural Coloration in Cephalopods."

126 **A multitude of pouches:** Justine J. Allen, George R. R. Bell, Alan M. Kuzirian, Sachin S. Velankar, and Roger T. Hanlon, "Comparative Morphology of Changeable Skin Papillae in Octopus and Cuttlefish," *Journal of Morphology* 275, no. 4 (April 2014): 371–90, doi.org/10.1002/jmor.20221.

Justine J. Allen, George R. R. Bell, Alan M. Kuzirian, and Roger T. Hanlon, "Cuttlefish Skin Papilla Morphology Suggests a Muscular Hydrostatic Function for Rapid Changeability," *Journal of Morphology* 274, no. 6 (June 2013): 645–56, doi.org/10.1002/jmor.20121.

126 **Cuttlefish blend into a variety:** Chuan-Chin Chiao and Roger T. Hanlon, "Rapid Adaptive Camouflage in Cephalopods," *Oxford Research Encyclopedia of Neuroscience*, December 2019, doi.org/10.1093/acrefore/9780190264086.013.182.

Theodosia Woo, Xitong Liang, Dominic A. Evans, Olivier Fernandez, Friedrich Kretschmer, Sam Reiter, and Gilles Laurent, "The Dynamics of Pattern Matching in Camouflaging Cuttlefish," *Nature* 619 (2023): 122–28, doi.org/10.1038/s41586-023-06259-2.

Montague, "Neural Control of Cephalopod Camouflage."

127 **"visual assessment shortcuts":** Chiao and Hanlon, "Rapid Adaptive Camouflage in Cephalopods."

127 **Beyond skin color and texture:** Alexandra Barbosa, Justine J. Allen, Lydia M. Mäthger, and Roger T. Hanlon, "Cuttlefish Use Visual Cues to Determine Arm Posture for Camouflage," *Proceedings of the Royal Society B* 279, no. 1726 (January 2012): 84–90, doi.org/10.1098/rspb.2011.0196.

127 **For example, the sand-dwelling:** Roger T. Hanlon, Anya C. Watson, and Alexandra Barbosa, "A 'Mimic Octopus' in the Atlantic: Flatfish Mimicry and Camouflage by *Macrotritopus defilippi*," *Biological Bulletin* 218, no. 1 (February 2010): 15–24, doi.org/10.1086/BBLv218n1p15.

128 **It not only matches:** Mark D. Norman, Julian Finn, and Tom Tregenza, "Dynamic Mimicry in an Indo-Malayan Octopus," *Proceedings of the Royal Society B* 268, no. 1478 (September 2001): 1755–58, doi.org/10.1098/rspb.2001.1708.

128 **Impersonating venomous or poisonous species:** Norman et al., "Dynamic Mimicry in an Indo-Malayan Octopus."

129 **"large-sized octopus is visible":** Aristotle, *Historia animalium*, book 4; D'Arcy Thompson, *A History of Animals* (Oxford: Clarendon Press, 1910).

130 **In 1819, Giosuè Sangiovanni:** Giosuè Sangiovanni, "Descrizione di un Particolare Sistema di Organi (Chromophoro-Espansivo-Dermoideo) e de' Fenomeni Ch'esso Produce; Scoverto ne' Molluschi Cefalopodi," *Giornale Enciclopedico di Napoli* 13, no. 3 (1819): 351–61.

NOTES

130 **In 1882, French naturalist:** Raphaël Blanchard, "The Chromatophores of Cephalopods," *Annals and Magazine of Natural History*, 6th ser., 9, no. 50 (1892): 182–83, doi.org/10.1080/00222939208677300.

130 **He thought that cephalopod chromatophores:** Societies and Academies, *Nature* 27 (March 1883): 473–76, doi.org/10.1038/027473b0.

130 **However, other researchers:** Césaer Phisalix, "On the Nature of the Movement of the Chromatophores of Cephalopods," *Annals and Magazine of Natural History*, 6th ser., 9, no. 50 (1892): 183–85, doi.org/10.1080/00222939208677301.

Paul Bert, "Mémoire sur la physiologie de la seiche," in *Notes d'anatomie et de physiologie comparées* (Paris: J.-B. Ballière, 1870), 49–72, doi.org/10.5962/bhl.title.102916.

Blanchard, "The Chromatophores of Cephalopods."

130 **Finally, in 1901, Eugen Steinach:** Eugen Steinach, "Studien über die Hautfärbung und über den Farbenwechsel der Cephalopoden," *Archiv für die gesamte Physiologie des Menschen und der Tiere* 87 (October 1901): 1–37, doi.org/10.1007/BF01657601.

130 **Eventually, in 1932, Enrico Sereni:** Enrico Sereni and J. Z. Young, "Nervous Degeneration and Regeneration in Cephalopods," *Pubblicazione della Stazione Zoologica di Napoli* 12 (1932): 173–208.

130 **In the late 1960s:** Richard A. Cloney and Ernst Florey, "Ultrastructure of Cephalopod Chromatophore Organs," *Zeitschrift für Zellforschung und mikroskopische Anatomie* 89 (June 1968): 250–80, doi.org/10.1007/BF00347297.

Ernst Florey, "Ultrastructure and Function of Cephalopod Chromatophores," *American Zoologist* 9, no. 2 (May 1969): 429–42, doi.org/10.1093/icb/9.2.429.

131 **Investigation of the other cell:** Siro Kawaguti and Shinji Ohgishi, "Electron Microscopic Study on Iridophores of a Cuttlefish, *Sepia esculenta*," *Biological Journal of Okayama University* 8, nos. 3–4 (December 1962): 115–29.

Eric James Denton and M. F. Land, "Mechanism of Relfexion in Silvery Layers of Fish and Cephalopods," *Proceedings of the Royal Society B* 178, no. 1050 (June 1971): 43–61, doi.org/10.1098/rspb.1971.0051.

NOTES

Steven L. Brocco, "The Fine Structure of the Frontal and Mantle White Spots of *Octopus Dofleini*," *American Society of Zoologists* 15, no. 3 (1975): 782.

131 **In squid, iridophores:** T. J. Wardill, P. T. Gonzalez-Bellido, R. J. Crook, and R. T. Hanlon, "Neural Control of Tuneable Skin Iridescence in Squid," *Proceedings of the Royal Society B* 279, no. 1745 (October 2012): 4243–52, doi.org/10.1098/rspb.2012.1374.

131 **Similarly, some squid:** Daniel G. DeMartini, Amitabh Ghoshal, Erica Pandolfi, Aaron T. Weaver, Mary Baum, and Daniel E. Morse, "Dynamic Biophotonics: Female Squid Exhibit Sexually Dimorphic Tunable Leucophores and Iridocytes," *Journal of Experimental Biology* 216, no. 19 (October 2013): 3733–41, doi.org/10.1242/jeb.090415.

132 **Smart materials are:** Mel Schwartz, preface to *Smart Materials,* ed. Mel Schwartz (Boca Raton, FL: CRC Press, 2009).

132 **Researchers from the University:** Aaron Fishman, Jonathan Rossiter, and Martin Homer, "Hiding the Squid: Patterns in Artificial Cephalopod Skin," *Journal of the Royal Society Interface* 12, no. 108 (July 2015): 20150281, doi.org/10.1098/rsif.2015.0281.

133 **He and his students produced:** J. H. Pikul, S. Li, H. Bai, R. T. Hanlon, I. Cohen, and R. F. Shepherd, "Stretchable Surfaces with Programmable 3D Texture Morphing for Synthetic Camouflaging Skins," *Science* 358, no. 6360 (October 2017): 210–14, doi.org/10.1126/science.aan5627.

133 **Their work with smart materials:** David J. Levine, Kevin T. Turner, and James H. Pikul, "Materials with Electroprogrammable Stiffness," *Advanced Materials* 33, no. 35 (September 2021): 2007952, doi.org/10.1002/adma.202007952.

CHAPTER 7: THE JELLYFISH'S LIGHT SHOW

144 **The Blaschka collection inspired:** David O. Brown, C. Drew Harvell, and Denis Jensen, *Fragile Legacy* (2015), fragilelegacy.info.

146 **They are distantly related:** Konstantin Khalturin, Chuya Shinzato, Maria Khalturina, Mayuko Hamada, Manabu Fujie, Ryo Koyanagi,

NOTES

Miyuki Kanda, et al., "Medusozoan Genomes Inform the Evolution of the Jellyfish Body Plan," *Nature Ecology & Evolution* 3 (2019): 811–22, doi.org/10.1038/s41559-019-0853-y.

Jun-Yuan Chen, Paola Oliveri, Feng Gao, Stephen Q. Dornbos, Chia-Wei Li, David J. Bottjer, and Eric H. Davidson, "Precambrian Animal Life: Probable Developmental and Adult Cnidarian Forms from Southwest China," *Developmental Biology* 248, no. 1 (August 2002): 182–96, doi.org/10.1006/dbio.2002.0714.

Paulyn Cartwright, Susan L. Halgedahl, Jonathan R. Hendricks, Richard D. Jarrard, Antonio C. Marques, Allen G. Collins, and Bruce L. Lieberman, "Exceptionally Preserved Jellyfishes from the Middle Cambrian," *PLoS ONE* 2, no. 10 (October 2007): e1121, doi.org/10.1371/journal.pone.0001121.

146 **Genetic differences among some:** Khalturin et al., "Medusozoan Genomes Inform the Evolution of the Jellyfish Body Plan."

148 **New work confirms the surprising:** Nathan V. Whelan, Kevin M. Kocot, Tatiana P. Moroz, Krishanu Mukherjee, Peter Williams, Gustav Paulay, Leonid L. Moroz, and Kenneth M. Halanych, "Ctenophore Relationships and Their Placement as the Sister Group to All Other Animals," *Nature Ecology & Evolution* 1, no. 11 (November 2017): 1737–46, doi.org/10.1038/s41559-017-0331-3.

Nathan V. Whelan, Kevin M. Kocot, Leonid L. Moroz, and Kenneth M. Halanych, "Error, Signal, and the Placement of Ctenophora Sister to All Other Animals," *PNAS* 112, no. 18 (April 2015): 5773–78, doi.org/10.1073/pnas.1503453112.

Kenneth M. Halanych, "The Ctenophore Lineage Is Older Than Sponges? That Cannot Be Right! Or Can It?," *Journal of Experimental Biology* 218, no. 2 (February 2015): 592–97, doi.org/10.1242/jeb.111872.

148 **First, large blocks of DNA:** Darrin T. Schultz, Steven H. D. Haddock, Jessen V. Bredeson, Richard E. Green, Oleg Simakov, and Daniel S. Rokhsar, "Ancient Gene Linkages Support Ctenophores as Sister to Other Animals," *Nature* 618 (2023): 110–17, doi.org/10.1038/s41586-023-05936-6.

148 **Instead of venom-filled nematocysts:** Kevin Pang and Mark Q. Martindale, "Ctenophores," *Current Biology* 18, no. 24 (December 2008): R1119–20.

NOTES

149 **Leopold Blaschka's words:** Henri Reiling, "The Blaschkas' Glass Animal Models: Origins of Design," *Journal of Glass Studies* 40 (1998): 105–26.

151 **This siphonophore jelly:** Casey Dunn, "Siphonophores," *Current Biology* 19, no. 6 (March 2009): R233–34, doi.org/10.1016/j.cub.2009.02.009.

151 **Large ones can reach:** "Giant Siphonophore," Monterey Bay Aquarium, accessed August 30, 2024, montereybayaquarium.org/animals/animals-a-to-z/giant-siphonophore.

152 **I've seen it reported:** E. A. Widder, S. A. Bernstein, D. F. Bracher, J. F. Case, K. R. Reisenbichler, J. J. Torres, and B. H. Robison, "Bioluminescence in the Monterey Submarine Canyon: Image Analysis of Video Recordings from a Midwater Submersible," *Marine Biology* 100, no. 4 (March 1989): 541–51, doi.org/10.1007/bf00394831.

152 **But they behave like:** G. O. Mackie, P. R. Pugh, and J. E. Purcell, "Siphonophore Biology," *Advances in Marine Biology* 24 (1988): 97–262, doi.org/10.1016/S0065-2881(08)60074-7.

152 **Haeckel was an intellectual:** Gregory S. Levit and Uwe Hossfeld, "Ernst Haeckel in the History of Biology," *Current Biology* 29, no. 24 (December 2019): R1276–84.

153 **His monograph on the siphonophores:** Ernst Haeckel, "Report on the Siphonophorae Collected by H. M. S. Challenger," in *Report on the Scientific Results of the Voyage of H. M. S. Challenger during the Years 1873–76*, ed. C. Wyville Thomson and John Murray (London: Eyre & Spottiswoode, 1888), doi.org/10.5962/bhl.title.6513.

153 **These are all in Cornell's:** "Out of the Teeming Sea: Cornell Collection of Blaschka Invertebrate Models," Cornell University Library, accessed August 30, 2024, digital.library.cornell.edu/collections/blaschka.

154 **This evolutionary ubiquity:** Steven H. D. Haddock, Mark A. Moline, and James F. Chase, "Bioluminescence in the Sea," *Annual Review of Marine Science* 2 (January 2010): 443–93, doi.org/10.1146/annurev-marine-120308-081028.

155 **For perspective, insects:** Y. Oba, K. Konishi, D. Yano, H. Shibata, D. Kato, T. T. Shirai, "Resurrecting the Ancient Glow of the Fireflies," *Science Advances* 6, no. 49 (December 2020), doi.org/10.1126/sciadv.abc5705.

NOTES

Gareth S. Powell, Natalie A. Saxon, Yelena M. Pacheco, Kathrin F. Stranger-Hall, Gavin J. Martin, Dominik Kusy, Luiz Felipe Lima Da Silveira, Ladislav Bocak, Marc A. Branham, and Seth M. Bybee, "Beetle Bioluminescence Outshines Extant Aerial Predators," *Proceedings of the Royal Society B* 289 (July 2022), doi.org/10.1098/rspb.2022.0821.

155 **Of all groups:** Séverine Martini and Steven H. D. Haddock, "Quantification of Bioluminescence from the Surface to the Deep Sea Demonstrates Its Predominance as an Ecological Trait," *Scientific Reports* 7 (2017): 45750, doi.org/10.1038/srep45750.

155 **Despite the ubiquity:** Chatragadda Ramesh and Manabu Bessho-Uehara, "Acquisition of Bioluminescent Trait by Non-luminous Organisms from Luminous Organisms through Various Origins," *Photochemical & Photobiological Sciences* 20, no. 11 (November 2021): 1547–62, doi.org/10.1007/s43630-021-00124-9.

155 **Dr. Kevin Raskoff of California State University:** "How the Jelly Got Its Glow," *Jellies Down Deep,* American Museum of Natural History, October 2011, video, 7:25, amnh.org/explore/videos/oceans/ocean-jellies.

156 **Bioluminescence is a predominant:** Haddock, Moline, and Chase, "Bioluminescence in the Sea."

156 **Researchers like Jim Morin:** James G. Morin, "'Firefleas' of the Sea: Luminescent Signaling in Marine Ostracode Crustaceans," *Florida Entomologist* 69, no. 1 (March 1986): 105–21, doi.org/10.2307/3494749.

Trevor J. Rivers and James G. Morin, "Complex Sexual Courtship Displays by Luminescent Male Marine Ostracods," *Journal of Experimental Biology* 211, no. 14 (July 2008): 2252–62, doi.org/10.1242/jeb.011130.

Trevor J. Rivers and James G. Morin, "Plasticity of Male Mating Behaviour in a Marine Bioluminescent Ostracod in Both Time and Space," *Animal Behaviour* 78, no. 3 (September 2009): 723–34, doi.org/10.1016/j.anbehav.2009.06.020.

James G. Morin, "Luminaries of the Reef: The History of Luminescent Ostracods and Their Courtship Displays in the Caribbean," *Journal of Crustacean Biology* 39, no. 3 (May 2019): 227–43, doi.org/10.1093/jcbiol/ruz009.

NOTES

157 **The viperfish (*Chauliodus sloani*):** Edith A. Widder, "Marine Bioluminescence: Why Do So Many Animals in the Open Ocean Make Light?," *Bioscience Explained* 1, no. 1 (2001).

157 **The deep-sea shrimp:** J. A. C. Nicol, "Observations on Luminescence in Pelagic Animals," *Journal of the Marine Biological Association of the United Kingdom* 37, no. 3 (October 1958): 705–52, doi.org/10.1017/S0025315400005749.

157 **Other animals flash light:** A. F. Mesinger and J. F. Case, "Dinoflagellate Luminescence Increases Susceptibility of Zooplankton to Teleost Predation," *Marine Biology* 112 (June 1992): 207–10, doi.org/10.1007/BF00702463.

Haddock, Moline, and Chase, "Bioluminescence in the Sea."

157 **When a luciferin reacts:** Osamu Shimomura, *Bioluminescence: Chemical Principles and Methods* (World Scientific Publishing, 2012).

158 **This activated compound:** J. W. Hastings and James G. Morin, "Calcium-Triggered Light Emission in Renilla. A Unitary Biochemical Scheme for Coelenterate Bioluminescence," *Biochemical and Biophysical Research Communications* 37, no. 3 (October 1969): 493–98, doi.org/10.1016/0006-291x(69)90942-5.

158 **All the jellies use:** Haddock, Moline, and Chase, "Bioluminescence in the Sea."

159 **Dr. Shimomura was obsessed:** Georgina Ferry, "Osamu Shimomura (1928–2018)," *Nature* 563, no. 7733 (November 2018): 627, doi.org/10.1038/d41586-018-07401-1.

Osamu Shimomura, "A Short Story of Aequorin," *Biological Bulletin* 189, no. 1 (August 1995): 1–5, doi.org/10.2307/1542194.

Osamu Shimomura, "The Discovery of Aequorin and Green Fluorescent Protein," *Journal of Microscopy* 217, no. 1 (January 2005): 3–15, doi.org/10.1111/j.0022-2720.2005.01441.x.

Osamu Shimomura, Frank H. Johnson, and Yo Saiga, "Extraction, Purification, and Properties of Aequorin, a Bioluminescent Protein from the Luminous Hydromedusan, *Aequorea*," *Journal of Cellular and Comparative Physiology* 59, no. 3 (June 1962): 223–39, doi.org/10.1002/jcp.1030590302.

NOTES

Osamu Shimomura, Frank H. Johnson, and Hiroshi Morise, "Mechanism of the Luminescent Intramolecular Reaction of Aequorin," *Biochemistry* 13, no. 16 (July 1974): 3278–86, doi.org/10.1021/bi00713a016.

Osamu Shimomura, "The Discovery of Aequorin and Green Fluorescent Protein," *Journal of Microscopy* 217, no. 1 (January 2005): 3–15.

160 **They need to eat seed:** Steven H. D. Haddock, Trevor J. Rivers, and Bruce H. Robison, "Can Coelenterates Make Coelenterazine? Dietary Requirement for Luciferin in Cnidarian Bioluminescence," *PNAS* 98, no. 20 (September 2001): 11148–51, doi.org/10.1073/pnas.201329798.

161 **He simply called it green:** Shimomura, Johnson, and Saiga, "Extraction, Purification and Properties of Aequorin."

161 **Jim was the one:** J. W. Hastings and J. G. Morin, "Comparative Biochemistry of Calcium-Activated Photoproteins from the Ctenophore, *Mnemiopsis* and the Coelenterates *Aequorea*, *Obelia*, *Pelagia* and *Renilla*," abstract, *Biological Bulletin* 137, no. 2 (October 1969), doi.org/10.1086/BBLv137n2p384.

Hiroshi Morise, Osamu Shimomura, Frank H. Johnson, and John Winant, "Intermolecular Energy Transfer in the Bioluminescent System of *Aequorea*," *Biochemistry* 13, no. 12 (June 1974): 2656–62, doi.org/10.1021/bi00709a028.

Osamu Shimomura, "Structure of the Chromophore of *Aequorea* Green Fluorescent Protein," *FEBS Letters* 104, no. 2 (August 1979): 220–22, doi.org/10.1016/0014-5793(79)80818-2.

Shimomura, "The Discovery of Aequorin and Green Fluorescent Protein."

161 **Two other scientists, Martin:** Martin Chalfie, "GFP: Lighting Up Life," *PNAS* 106, no. 25 (June 2009): 10073–80, doi.org/10.1073/pnas.0904061106.

Roger Y. Tsien, "Nobel Lecture," Constructing and Exploiting the Fluorescent Protein Paintbox," *Integrative Biology* 2, no. 2–3 (March 2010): 77–93, doi.org/10.1039/B926500G.

161 **Together in 2008:** Paul S. Weiss, "2008 Nobel Prize in Chemistry: Green Fluorescent Protein, Its Variants and Implications," *ACS Nano* 2, no. 10 (October 2008): 1977, doi.org/10.1021/nn800671h.

NOTES

Osamu Shimomura, "Discovery of Green Fluorescent Protein (GFP) (Nobel Lecture)," *Angewandte Chemie* 48, no. 31 (July 2009): 5590–602, doi.org/10.1002/anie.200902240.

162 **Because the brightness:** A. B. Borle and K. W. Snowdowne, "Measurement of Intracellular Free Calcium in Monkey Kidney Cells with Aequorin," *Science* 217, no. 4556 (July 1982): 252–54, doi.org/10.1126/science.6806904.

Massimo Bonora, Carlotta Giorgi, Angela Bononi, Saverio Marchi, Simone Patergnani, Alessandro Rimessi, Rosario Rizzuto, and Paolo Pinton, "Subcellular Calcium Measurements in Mammalian Cells Using Jellyfish Photoprotein Aequorin-Based Probes," *Nature Protocols* 8 (2013): 2105–18, doi.org/10.1038/nprot.2013.127.

J. Alvarez and M. Montero, "Measuring [Ca^{+2}] in the Endoplasmic Reticulum with Aequorin," *Cell Calcium* 32, nos. 5–6 (November–December 2002): 251–60, doi.org/10.1016/S0143416002001860.

162 **The level of calcium:** David E. Clapham, "Calcium Signaling," *Cell* 131, no. 6 (December 2007): 1047–58, doi.org/10.1016/j.cell.2007.11.028.

162 **The measurable brightness of aequorin:** Carlotta Giorgi, Massimo Bonora, Sonia Missiroli, Federica Poletti, Fabian Galindo Ramirez, Giampaolo Morciano, Claudia Morganti, Pier Paolo Pandolfi, Fabio Mammano, and Paolo Pinton, "Intravital Imaging Reveals p53-Dependent Cancer Cell Death Induced by Phototherapy via Calcium Signaling," *Oncotarget* 6, no. 3 (January 2015): 1435–45, doi.org/10.18632/oncotarget.2935.

Luísa Cortes, João Malva, Ana Cristina Rego, and Cláudia F. Pereira, "Calcium Signaling in Aging and Neurodegenerative Diseases 2019," *International Journal of Molecular Sciences* 21, no. 3 (February 2020): 1–7, doi.org/10.3390/ijms21031125.

Fulvio Celsi, Paolo Pizzo, Marisa Brini, Sara Leo, Carmen Fotino, Paolo Pinton, and Rosario Rizzuto, "Mitochondria, Calcium and Cell Death: A Deadly Triad in Neurodegeneration," *Biochimica et Biophysica Acta (BBA)—Bioenergetics* 1787, no. 5 (May 2009): 335–44, doi.org/10.1016/j.bbabio.2009.02.021.

163 **For example, in transgenic organisms:** C. Neal Stewart Jr., "Go with the Glow: Fluorescent Proteins to Light Transgenic Organisms," *Trends in*

NOTES

Biotechnology 24, no. 4 (April 2006): 155–62, doi.org/10.1016/j.tibtech.2006.02.002.

163 **GFP can also tag:** Robert M. Hoffman, "Application of GFP Imaging in Cancer," *Laboratory Investigation* 95, no. 4 (April 2015): 432–52, doi.org/10.1038/labinvest.2014.154.

163 **GFP is widely used:** Zexun Lu, Riccardo Tombolini, Sheridan Woo, Susanne Zeilinger, Matteo Lorito, and Janet K. Jansson, "In Vivo Study of *Trichoderma*-Pathogen-Plant Interactions, Using Constitutive and Inducible Green Fluorescent Protein Reporter Systems," *Applied and Environmental Microbiology* 70, no. 5 (May 2004): 3073–81, doi.org/10.1128/AEM.70.5.3073-3081.2004.

F. Joseph Pollock, Cory J. Krediet, Melissa Garren, Roman Stocker, Karina Winn, Bryan Wilson, Carla Huete-Stauffer, Bette L. Willis, and David G. Bourne, "Visualization of Coral Host-Pathogen Interactions Using a Stable GFP-Labeled *Vibrio coralliilyticus* Strain," *Coral Reefs* 34, no. 2 (June 2015): 655–62, doi.org/10.1007/s00338-015-1273-3.

CHAPTER 8: THE SEA STAR'S STICKY SKIN

167 **Unexpected as it seems:** John M. Lawrence, *Biology and Ecology of the Asteroidea* (Baltimore: John Hopkins University Press, 2013).

168 **Sea stars have little ability:** William B. Stickle and Walter J. Diehl, "Effects of Salinity on Echinoderms," in *Echinoderm Studies,* vol. 2, ed. Michel Jangoux and John M. Lawrence (Rotterdam: A. A. Balkema, 1987).

168 **It can change to be:** I. C. Wilkie, M. Sugni, H. S. Gupta, M. D. Candia Carnevali, and M. R. Elphick, "The Mutable Collagenous Tissue of Echinoderms: From Biology to Biomedical Applications," in *Soft Matter for Biomedical Applications*, ed. Helena S. Azevedo, João F. Mano, and João Borges (Royal Society of Chemistry, 2021), 1–33, doi.org/10.1039/9781839161124-00001.

168 **They are all extremely ancient:** Samuel Zamora and Imran A. Rahman, "Deciphering the Early Evolution of Echinoderms with Cambrian Fossils," *Frontiers in Paleontology* 57, no. 6 (November 2014): 1105–19, doi.org/10.1111/pala.12138.

NOTES

169 **We are both deuterostomes:** Claus Nielsen, "Evolution of Deuterostomy—and Origin of the Chordates," *Biological Reviews* 92, no. 1 (October 2015): 316–25, doi.org/10.1111/brv.12229.

169 **Bob claimed this star:** R. T. Paine, "A Note on Trophic Complexity and Community Stability," *American Naturalist* 103, no. 929 (January–February 1969): 91–93, doi.org/10.1086/282586.

170 **But Bob was brilliant:** Robert T. Paine, "Intertidal Community Structure," *Oecologia* 15 (June 1974): 93–120, doi.org/10.1007/BF00345739.

170 **He mathematically defined a keystone consumer:** Robert T. Paine, "Food Chain Dynamics and Trophic Cascades in Intertidal Habitats," in *Trophic Cascades: Predators, Prey, and the Changing Dynamics of Nature*, ed. John Terborgh and James A. Estes (Washington, DC: Island Press, 2010).

173 **Bob coined the term:** Paine, "A Note on Trophic Complexity."

174 **The sea star's superpower is:** Marcel E. Lavoie, "How Sea Stars Open Bivalves," *Biological Bulletin* 111, no. 1 (August 1956): 114–22, doi.org/10.2307/1539188.

174 **sea star always wins:** Lavoie, "How Sea Stars Open Bivalves."

175 **Jacques Amand Eudes-Deslongchamps:** Jacques Amand Eudes-Deslongchamps, "Notes sur l'Astérie commune," *Annales des sciences naturelles* 9 (1826): 219–21.

175 **As recently as 1932:** Eisiro Sawano and Kinji Mitsugi, "Toxic Action of the Stomach Extracts of the Starfishes on the Heart of the Oyster," *Science Reports of the Tohoku Imperial University*, 4th ser. (Biology) 7, no. 1 (1932): 79–88.

175 **The other theory, that great:** Paul Fischer, "Faune conchyliologique marine du Département de la Gironde et des côtes du sud-ouest de la France," *Actes de la Société Linnéenne de Bordeaux* 5 (1864): 257–338, doi.org/10.5962/bhl.title.13109.

Francis Jeffrey Bell, *Catalogue of the British Echinoderms in the British Museum* (London: Taylor and Francis, 1892).

175 **In 1895, Paulus Schiemenz:** Paulus Schiemenz, "Wie offen die Seestern Austern?," *Mittheilungen des Deutschen Seefischereivereins* 12, no. 6 (1895): 102–18.

NOTES

176 **Marcel Lavoie tested both theories:** Lavoie, "How Sea Stars Open Bivalves."

177 **Sea star skin is spiny:** Patricia O'Neill, "Structure and Mechanics of Starfish Body Wall," *Journal of Experimental Biology* 147, no. 1 (November 1989): 53–89, doi.org/10.1242/jeb.147.1.53.

177 **A stiff posture allows:** Tatsuo Motokawa, Eriko Santo, and Kenichi Umeyama, "Energy Expenditure Associated with Softening and Stiffening of Echinoderm Connective Tissue," *Biological Bulletin* 222, no. 2 (April 2012): 150–57, doi.org/10.1086/BBLv222n2p150.

177 **The real prize is that:** Iain C. Wilkie, "Is Muscle Involved in the Mechanical Adaptability of Echinoderm Mutable Collagenous Tissue?," *Journal of Experimental Biology* 205, no. 2 (January 2002): 159–65, doi.org/10.1242/jeb.205.2.159.

177 **They found that alterations:** Greg K. Szulgit and Robert E. Shadwick, "Dynamic Mechanical Characterization of a Mutable Collagenous Tissue: Response of Sea Cucumber Dermis to Cell Lysis and Dermal Extracts," *Journal of Experimental Biology* 203, no. 10 (May 2000): 1539–50, doi.org/10.1242/jeb.203.10.1539.

178 **Regeneration is common:** Iain C. Wilkie, "Autotomy as a Prelude to Regeneration in Echinoderms," *Microscopy Research and Technique* 55, no. 6 (December 2001): 369–96, doi.org/10.1002/jemt.1185.

178 **The autotomy process:** Iain C. Wilkie and M. Daniela Candia Carnevali, "Morphological and Physiological Aspects of Mutable Collagenous Tissue at the Autotomy Plane of the Starfish *Asterias rubens* L. (Echinodermata, Asteroidea): An Echinoderm Paradigm," *Marine Drugs* 21, no. 3 (2023): 138, doi.org/10.3390/md21030138.

178 **Some species have taken:** P. V. Mladenov and R. D. Burke, "Echinodermata: Asexual Propagation," in *Reproductive Biology of Invertebrates*, ed. K. G. Adiyodi and R. G. Adiyodi, vol. 6, part B, *Asexual Propagation and Reproductive Strategies* (New Delhi: Oxford and IBD, 1994), 339–83.

179 **The smart skin of sea stars:** Wilkie et al., "The Mutable Collagenous Tissue of Echinoderms."

Mel Schwartz, preface to *Smart Materials*, ed. Mel Schwartz (Boca Raton, FL: CRC Press, 2009).

NOTES

J. A. Trotter, J. Tipper, G. Lyons-Levy, K. Chino, A. H. Heuer, Z. Liu, M. Mrkisch, et al., "Towards a Fibrous Composite with Dynamically Controlled Stiffness: Lessons from Echinoderms," *Biochemical Society Transactions* 28, no. 4 (August 2000): 357–62, doi.org/10.1042/bst0280357.

S. J. Kew, J. H. Gwynne, D. Enea, M. Abu-Rub, A. Pandit, D. Zeugolis, R. A. Brooks, N. Rushton, S. M. Best, and R. E. Cameron, "Regeneration and Repair of Tendon and Ligament Tissue Using Collagen Fibre Biomaterials," *Acta Biomaterialia* 7, no. 9 (September 2011): 3237–47, doi.org/10.1016/j.actbio.2011.06.002.

Jingyi Mo, Sylvain F. Prévost, Liisa M. Blowes, Michaela Egertová, Nicholas J. Terrill, Wen Wang, Maurice R. Elphick, and Himadri S. Gupta, "Interfibrillar Stiffening of Echinoderm Mutable Collagenous Tissue Demonstrated at the Nanoscale," *PNAS* 113, no. 42 (October 2016): E6362–71, doi.org/10.1073/pnas.1609341113.

179 **They designed a hybrid biomaterial:** Trotter et al., "Towards a Fibrous Composite with Dynamically Controlled Stiffness."

179 **Development of new synthetic:** Kew et al., "Regeneration and Repair of Tendon and Ligament Tissue."

180 **Understanding the molecular mechanisms:** Mo et al., "Interfibrillar Stiffening of Echinoderm Mutable Collagenous Tissue."

181 **The onset of this scale:** Ian Hewson, Jason B. Button, Brent M. Gudenkauf, Benjamin Miner, Alisa L. Newton, Joseph K. Gaydos, Janna Wynne, et al., "Densovirus Associated with Sea-Star Wasting Disease and Mass Mortality," *PNAS* 111, no. 48 (December 2014): 17278–83, doi.org/10.1073/pnas.1416625111.

Diego Montecino-Latorre, Morgan E. Eisenlord, Margaret Turner, Reyn Yoshioka, C. Drew Harvell, Christy V. Pattengill-Semmens, Janna D. Nichols, and Joseph K. Gaydos, "Devastating Transboundary Impacts of Sea Star Wasting Disease on Subtidal Asteroids," *PLoS ONE* 11, no. 10 (2016): e0163190, doi.org/10.1371/journal.pone.0163190.

182 **A PBS newscast:** PBS NewsHour, "Mysterious Epidemic Devastates Starfish Populations off the Pacific Coast," PBS, January 30, 2014, pbs.org/newshour/show/mysterious-epidemic-devastates-starfish-population-pacific-coast#transcript.

183 **Thousands of diver surveys:** C. Drew Harvell, D. Montecino-Latorre, J. M. Caldwell, J. M. Burt, K. Bosley, A. Keller, S. F. Heron, et al., "Disease Epidemic and a Marine Heat Wave Are Associated with the Continental-Scale Collapse of a Pivotal Predator (*Pycnopodia helianthoides*)," *Science Advances* 5, no. 1 (January 2019): doi.org/10.1126/sciadv.aau7042.

> S.A. Gravem, W. N. Heady, V. R. Saccomanno, K. F. Alvstad, A. L. M. Gehman, T. N. Frierson, and S. L. Hamilton, "Sunflower Sea Star (*Pycnopodia helianthoides*)," *IUCN Red List Assessment* (2021).

185 **Each tube foot is adhesive:** Elise Hennebert, Pascal Viville, Roberto Lazzaroni, and Patrick Flammang, "Micro- and Nanostructure of the Adhesive Material Secreted by the Tube Feet of the Sea Star *Asterias rubens*," *Journal of Structural Biology* 164, no. 1 (October 2008): 108–18, doi.org/10.1016/j.jsb.2008.06.007.

EPILOGUE: SPINELESS FUTURES IN A WARMING OCEAN

190 **They noticed salmon:** Gideon J. Mordecai, Kristina M. Miller, Emiliano Di Cicco, Angela D. Schulze, Karia H. Kaukinen, Tobi J. Ming, Shaorong Li, et al., "Endangered Wild Salmon Infected by Newly Discovered Viruses," *eLife* 8 (2019): e47615, doi.org/10.7554/eLife.47615.

190 **What can we learn:** C. Drew Harvell, "How Starfish, Snails and Salmon Fight Pandemics," *New York Times*, April 17, 2020, nytimes.com/2020/04/17/opinion/covid-ocean-research.html.

191 **In fifty years, the oceans:** Alexander T. Lowe, Julia Bos, and Jennifer Ruesink, "Ecosystem Metabolism Drives pH Variability and Modulates Long-Term Ocean Acidification in the Northeast Pacific Coastal Ocean," *Scientific Reports* 9 (2019): 963, doi.org/10.1038/s41598-018-37764-4.

> Nicholas R. Bates, Yrene M. Astor, Matthew J. Church, Kim Currie, John E. Dore, Melchor González-Dávila, Laura Lorenzoni, Frank Muller-Karger, Jon Olafsson, and J. Magdalena Santana-Casiano, "A Time-Series View of Changing Surface Ocean Chemistry Due to Ocean Uptake of Anthropogenic CO_2 and Ocean Acidification,"

Oceanography 27, no. 1 (March 2014): 126–41, doi.org/10.5670/oceanog.2014.16.

Scott C. Doney, Victoria J. Fabry, Richard A. Feely, and Joan A. Kleypas, "Ocean Acidification: The Other CO_2 Problem," *Annual Review of Marine Science* 1 (2009): 169–92, doi.org/10.1146/annurev.marine.010908.163834.

191 **The seasonally low-pH waters:** D. Shallin Busch, Michael Maher, Patricia Thibodeau, and Paul McElhany, "Shell Condition and Survival of Puget Sound Pteropods Are Impaired by Ocean Acidification Conditions," *PLoS ONE* 9, no. 8 (August 2014): e105884, doi.org/10.1371/journal.pone.0105884.

Richard A. Feely, Simone R. Alin, Jan Newton, Christopher L. Sabine, Mark Warner, Allan Devol, Christopher Krembs, and Carol Maloy, "The Combined Effects of Ocean Acidification, Mixing, and Respiration on pH and Carbonate Saturation in an Urbanized Estuary," *Estuarine, Coastal and Shelf Science* 88, no. 4 (August 2010): 442–49, doi.org/10.1016/j.ecss.2010.05.004.

Alan Barton, George G. Waldbusser, Richard A. Feely, Stephen B. Weisberg, Jan A. Newton, Burke Hales, Sue Cudd, Benoit Eudeline, Chris J. Langdon, and Ian Jefferds, et al., "Impacts of Coastal Acidification on the Pacific Northwest Shellfish Industry and Adaptation Strategies Implemented in Response," *Oceanography* 28, no. 2 (June 2015): 146–59, doi.org/10.5670/oceanog.2015.38.

Sarah R. Cooley, Jack E. Cheney, Ryan P. Kelly, and Edward H. Allison, "Ocean Acidification and Pacific Oyster Larval Failures in the Pacific Northwest United States," in *Global Change in Marine Systems*, ed. Patrice Guillotreau, Alida Bundy, and R. Ian Perry (New York: Routledge, 2018).

191 **The current rate of warming:** IPCC, Climate Change 2023: Synthesis Report, ed. Core Writing Team, Hoesung Lee, and José Romero (Geneva, Switzerland: IPCC, 2023), 35–115, doi.org/10.59327/IPCC/AR6-9789291691647.

Edward Blanchard Wrigglesworth, Tyler Cox, Zachary I. Espinosa, and Aaron Donohoe, "The Largest Ever Recorded Heatwave—Characteristics and Attribution of the Antarctic Heatwave of

March 2022," *Geophysical Research Letters* 50, no. 17 (September 2023): doi.org/10.1029/2023GL104910.

193 **By 2008, following multiple:** Kent E. Carpenter, Muhammad Abrar, Greta Aeby, Richard B. Aronson, Stuart Banks, Andrew Bruckner, Angel Chiriboga, et al., "One-Third of Reef-Building Corals Face Elevated Extinction Risk from Climate Change and Local Impacts," *Science* 321, no. 5888 (July 2008): 560–63, doi.org/10.1126/science.1159196.

193 **While coral reefs the world:** Danielle Hall, "An Ocean Heat Wave Reaches to the Seafloor," Smithsonian National Museum of Natural History, August 2023, ocean.si.edu/conservation/climate-change/ocean-heat-wave-reaches-seafloor.

194 **Along with the heat stress:** Neil C. S. Chan and Sean R. Connolly, "Sensitivity of Coral Calcification to Ocean Acidification: A Meta-Analysis," *Global Change Biology* 19, no. 1 (January 2013): 282–90, doi.org/10.1111/gcb.12011.

Jonathan Erez, Stéphanie Reynaud, Jacob Silverman, Kenneth Schneider, and Denis Allemand, "Coral Calcification under Ocean Acidification and Global Change," in *Coral Reefs: An Ecosystem in Transition*, ed. Zvy Dubinsky and Noga Stambler (Dordrecht, Netherlands: Springer, 2010), 151–76, doi.org/10.1007/978-94-007-0114-4_10.

Niklas A. Kornder, Bernhard M. Riegl, and Joana Figueiredo, "Thresholds and Drivers of Coral Calcification Reponses to Climate Change," *Global Change Biology* 24, no. 11 (November 2018): 5084–95, doi.org/10.1111/gcb.14431.

David I. Kline, Lida Teneva, Daniel K. Okamoto, Kenneth Schneider, Ken Caldeira, Thomas Miard, Aaron Chai, et al., "Living Coral Tissue Slows Skeletal Dissolution Related to Ocean Acidification," *Nature Ecology & Evolution* 3 (2019): 1438–44, doi.org/10.1038/s41559-019-0988-x.

194 **Our work showed that their immune:** Laura D. Mydlarz, Sally F. Holthouse, Esther C. Peters, and C. Drew Harvell, "Cellular Responses in Sea Fan Corals: Granular Amoebocytes React to Pathogen and Climate Stressors," *PLoS ONE* 3, no. 3 (March 2008): e1811, doi.org/10.1371/journal.pone.0001811.

NOTES

195 **A new study that asks:** Mary Alice Coffroth, Louis A. Buccella, Katherine M. Eaton, Howard R. Lasker, Alyssa T. Gooding, and Harleena Franklin, "What Makes a Winner? Symbiont and Host Dynamics Determine Caribbean Octocoral Resilience to Bleaching," *Science Advances* 9, no. 47 (November 2023): eadj6788, doi.org/10.1126/sciadv.adj6788.

196 **Barrel sponges increased:** Steven E. McMurray, Timothy P. Henkel, and Joseph P. Pawlik, "Demographics of Increasing Populations of the Giant Barrel Sponge *Xestospongia muta* in the Florida Keys," *Ecology* 91, no. 2 (February 2010): 560–70, doi.org/10.1890/08-2060.1.

196 **There have been news:** Priya Shukla, "As Oceans Warm, Jellyfish Swarm," *Forbes*, October 13, 2021, forbes.com/sites/priyashukla/2021/10/13/as-oceans-warm-jellyfish-swarm/?sh=701c179e4731.

Aylin Woodward, "Thousands of Animals around the World Are at Risk of Extinction. But Not Jellyfish—They're Thriving in Warm, Polluted Water," *Business Insider*, October 30, 2019, businessinsider.com/jellyfish-thriving-climate-change-warm-oceans-2019-10.

Susan D'Agostino, "Jellyfish Attack Nuclear Power Plant. Again," *Bulletin of the Atomic Scientists*, October 28, 2021, thebulletin.org/2021/10/jellyfish-attack-nuclear-power-plant-again.

196 **You would think from these:** Juliet Lamb, "The Global Jellyfish Crisis in Perspective," *JSTOR Daily,* March 29, 2017, daily.jstor.org/global-jellyfish-crisis-perspective.

Robert H. Codon, William M. Graham, Carlos M. Duarte, Kylie A. Pitt, Cathy H. Lucas, Steven H. D. Haddock, Kelly R. Sutherland, et al., "Questioning the Rise of Gelatinous Zooplankton in the World's Oceans," *BioScience* 62, no. 2 (February 2012): 160–69, doi.org/10.1525/bio.2012.62.2.9.

Robert H. Condon, Carlos M. Duarte, Kylie A. Pitt, Kelly L. Robinson, Cathy H. Lucas, Kelly R. Sutherland, Hermes W. Miazan, et al., "Recurrent Jellyfish Blooms Are a Consequence of Global Oscillations," *PNAS* 110, no. 3 (January 2012): 1000–1005, doi.org/10.1073/pnas.1210920110.

196 **However, there are many jelly:** Claudia E. Mills, "Jellyfish Blooms: Are Populations Increasing Globally in Response to Changing Ocean

Conditions?," *Hydrobiologia* 451, no. 1–3 (May 2001): 55–68, doi.org /10.1023/a:1011888006302.

Sabine Holst, "Effects of Climate Warming on Strobilation and Ephyra Production of North Sea Scyphozoan Jellyfish," in *Jellyfish Blooms IV*, ed. Jennifer Purcell, Hermes Miazan, and Jessica R. Frost (Dordrecht, Netherlands: Springer, 2012), 127–40, doi.org /10.1007/978-94-007-5316-7_10.

197 **Scientists have shown that coral:** Caitlyn Kennedy, "Ocean Acidification, Today and in the Future," Climate.gov, NOAA, November 3, 2010, climate.gov/news-features/featured-images/ocean-acidification -today-and-future.

197 **Studies show that in California:** Stewart T. Schultz, Jeffrey H. R. Goddard, Terrence M. Gosliner, Douglas E. Mason, William E. Pence, Gary R. McDonald, Vicki B. Pearse, and John S. Pearse, "Climate-Index Response Profiling Indicates Larval Transport Is Driving Population Fluctuations in Nudibranch Gastropods from the Northeast Pacific Ocean," *Limnology and Oceanography* 56, no. 2 (March 2011): 749–63, doi.org/10.4319/lo.2011.56.2.0749.

197 **As squid and octopuses have:** Zoë A. Doubleday, Thomas A. A. Prowse, Alexander Arkhipkin, Graham J. Pierce, Jayson Semmens, Michael Steer, Stephen C. Leporati, et al., "Global Proliferation of Cephalopods," *Current Biology* 26, no. 10 (May 2016): R406–7, doi .org/10.1016/j.cub.2016.04.002.

Sean C. Anderson, Joanna Mills Flemming, Reg Watson, and Heike K. Lotze, "Rapid Global Expansion of Invertebrate Fisheries: Trends, Drivers, and Ecosystem Effects," *PLoS ONE* 6, no. 3 (March 2011): e14735, doi.org/10.1371/journal.pone.0014735.

198 **We do know that the:** M. Vargas-Yáñez, F. Moya, M. García-Martínez, J. Rey, M. González, and P. Zunino, "Relationships between *Octopus vulgaris* Landings and Environmental Factors in the Northern Alboran Sea (Southwestern Mediterranean)," *Fisheries Research* 99, no. 3 (August 2009): 159–67, doi.org/10.1016/j.fishres.2009.05.013.

Tiago Repolho, Miguel Baptista, Marta S. Pimentel, Gisela Dionísio, Katja Trübenbach, Vanessa M. Lopes, Ana Rita Lopes, Ricardo Calado, Mário Diniz, and Rui Rosa, "Developmental and Physiological

Challenges of Octopus (*Octopus vulgaris*) Early Life Stages under Ocean Warming," *Journal of Comparative Physiology B* 184, no. 1 (January 2014): 55–64, doi.org/10.1007/s00360-013-0783-y.

Francisco Oliveira Borges, Miguel Guerreiro, Catarina Pereira Santos, José Ricardo Paula, and Rui Rosa, "Projecting Future Climate Change Impacts on the Distribution of the '*Octopus vulgaris* Species Complex,'" *Frontiers in Marine Science* 9 (2022): 1018766, doi.org/10.3389/fmars.2022.1018766.

198 **When stressed, these clams:** J. H. Norton, H. C. Prior, B. Baillie, and D. Yellowlees, "Atrophy of the Zooxanthellal Tubular System in Bleached Giant Clams *Tridacna gigas*," *Journal of Invertebrate Pathology* 66, no. 3 (November 1995): 307–10, doi.org/10.1006/jipa.1995.1106.

Bela Hieronymus Buck, Harald Rosenthal, and Ulrich Saint-Paul, "Effect of Increased Irradiance and Thermal Stress on the Symbiosis of *Symbiodinium microadriaticum* and *Tridacna gigas*," *Aquatic Living Resources* 15, no. 2 (April 2002): 107–17, doi.org/10.1016/S0990-7440(02)01159-2.

198 **Climate change also acts indirectly:** Jun Li, Yinyin Zhou, Yanpin Qin, Jinkuan Wei, Pengyang Shigong, Haitao Ma, Yunqing Li, et al., "Assessment of the Juvenile Vulnerability of Symbiont-Bearing Giant Clams to Ocean Acidification," *Science of the Total Environment* 812 (March 2022): 152265, doi.org/10.1016/j.scitotenv.2021.152265.

198 **Greater acidification itself causes higher:** Sue-Ann Watson, Paul C. South-Gate, Gabrielle M. Miller, Jonathan A. Moorhead, and Jens Knauer, "Ocean Acidification and Warming Reduce Juvenile Survival of the Fluted Giant Clam, *Tridacna squamosa*," *Molluscan Research* 32, no. 3 (September 2012): 177–80, doi.org/10.11646/mr.32.3.7.

199 **The decline of sunflower stars:** Drew Harvell, "The Domino Effect of an Underwater Disease Outbreak," *The Hill*, September 19, 2014, thehill.com/blogs/pundits-blog/energy-environment/218234-the-domino-effect-of-an-underwater-disease-outbreak.

199 **There are still many questions:** Amanda E. Bates, Brett J. Hilton, and Christopher D. G. Harley, "Effects of Temperature, Season and Locality on Wasting Disease in the Keystone Predatory Sea Star *Pisaster ochraceus*," *Diseases of Aquatic Organisms* 86, no. 3 (2009): 245–51, doi.org/10.3354/dao02125.

199 **Slack skin and rapid loss:** Ian Hewson, Jason B. Button, Brent M. Gudenkauf, Benjamin Miner, Alisa L. Newton, Joseph K. Gaydos, Janna Wynne, et al., "Densovirus Associated with Sea-Star Wasting Disease and Mass Mortality," *PNAS* 111, no. 48 (December 2014): 17278–83, doi.org/10.1073/pnas.1416625111.

201 **Other studies show MPAs:** Alexandra Smith, Juan Domingo Aguilar, Charles Boch, Guilio De Leo, Arturo Hernández-Velasco, Stephanie Houck, Ramón Martinez, Stephen Monismith, Jorge Torre, C. Brock Woodson, et al., "Rapid Recovery of Depleted Abalone in Isla Natividad, Baja California, Mexico," *Ecosphere* 13, no. 3 (March 2022): e4002, doi.org/10.1002/ecs2.4002.

Joleah B. Lamb, David H. Williamson, Garry R. Russ, and Bette L. Willis, "Protected Areas Mitigate Diseases of Reef-Building Corals by Reducing Damage from Fishing," *Ecology* 96, no. 9 (September 2015): 2555–67, doi.org/10.1890/14-1952.1.

202 **Dr. Victor Bonito works:** Lice Monovo and Nick Sas, "Fiji's Beaches and Reefs Face an Uncertain Future but This Scientist Says 'Assisted Evolution' Can Keep Them Alive," ABC News, July 1, 2023, abc.net.au/news/2023-07-02/assisted-evolution-to-keep-fiji-reefs-alive/102507974.

203 **In addition to whole ecosystem:** Hanny E. Rivera, Anne L. Cohen, Janelle R. Thompson, Iliana B. Baums, Michael D. Fox, and Kristin S. Meyer-Kaiser, "Palau's Warmest Reefs Harbor Thermally Tolerant Corals That Thrive across Different Habitats," *Communications Biology* 5 (2022): 1394, doi.org/10.1038/s42003-022-04315-7.

Line K. Bay, Juan C. Ortiz, Adriana Humanes, Cynthia Riginos, Iliana B. Baums, Hugo Scharfenstein, Manuel Aranda, et al., *R&D Technology Roadmap for Understanding Natural Adaptation and Assisted Evolution of Corals to Climate Change.* CORDAP (2023).

203 **There are three main components:** Talisa Doering, Justin Maire, Madeleine J. H. van Oppen, and Linda L. Blackall, "Advancing Coral Microbiome Manipulation to Build Long-Term Climate Resilience," *Microbiology Australia* 44, no. 1 (2023): 36–40, doi.org/10.1071/MA23009.

Wing Yan Chan, Luka Meyers, David Rudd, Sanjida H. Topa, and Madeleine J. H. van Oppen, "Heat-Evolved Algal Symbionts Enhance Bleaching Tolerance of Adult Corals without Trade-Off against

Growth," *Global Change Biology* 29, no. 24 (December 2023): 6945–68, doi.org/10.1111/gcb.16987.

Lochan Chaudhari and Charvi Trivedi, "Inducing Heat Tolerance in Corals Using Genetic Modification in Host and Symbiont Simultaneously," *ECS Transactions* 107 (2022): 14655, doi.org/10.1149/10701.14655ecst.

204 **In addition to the well-known:** Joleah B. Lamb, Jeroen A. J. M. Van de Water, David G. Bourne, Craig Altier, Margaux Y. Hein, Evan A. Fiorenza, Nur Abu, Jamaluddin Jompa, and C. Drew Harvell, "Seagrass Ecosystems Reduce Exposure to Bacterial Pathogens of Humas, Fishes, and Invertebrates," *Science* 355, no. 6326 (February 2017): 731–33, doi.org/10.1126/science.aal1956.

Phoebe Dawkins, Evan A. Fiorenza, Jeffrey Gaeckle, Jennifer Lanksbury, C. Drew Harvell, and Joleah B. Lamb, "Seagrasses Mitigate Human Pathogens in Shellfish along an Urban Gradient" (Ecological Society of America Meeting, August 2020).

204 **But, like coral reefs, seagrass:** Lillian R. Aoki, Bo Yang, Olivia J. Graham, Carla Gomes, Brendan Rappazzo, Timothy L. Hawthorne, J. Emmett Duffy, and Drew Harvell, "UAV High-Resolution Imaging and Disease Surveys Combine to Quantify Climate-Related Decline in Seagrass," *Frontiers in Ocean Observing* 36, no. 1 (2023): 38–39, doi.org/10.5670/oceanog.2023.s1.12.

Olivia J. Graham, Lillian R. Aoki, Tiffany Stephens, Joshua Stokes, Sukanya Dayal, Brendan Rappazzo, Carla P. Gomes, and C. Drew Harvell, "Effects of Seagrass Wasting Disease on Eelgrass Growth and Belowground Sugar in Natural Meadows," *Frontiers in Marine Science* 8 (November 2021): 768668, doi.org/10.3389/fmars.2021.768668.

205 **The goal of the National Nature Assessment:** U.S. Global Change Research Program, "National Nature Assessment," globalchange.gov/our-work/national-nature-assessment.

206 **"Nature-based solutions should be go-to":** Heather Tallis, "Nature-Based Solutions from the Halls of the Exec Office of the President," interview by Sarah Thorne and Todd Bridges, December 14, 2022, in *The Engineering with Nature Podcast*, 37:27, produced by iContact

NOTES

Productions, ewn.erdc.dren.mil/podcasts/episode/s5-e2-nature-based-solutions-from-the-halls-of-the-exec-office-of-the-president.

207 **Amid a flurry of biomedical:** Rodolphe Barrangou and Philippe Horvath, "A Decade of Discovery: CRISPR Functions and Applications," *Nature Microbiology* 2 (2017): 17092, doi.org/10.1038/nmicrobiol.2017.92.

207 **It was discovered in 2007:** Rodolphe Barrangou, Christophe Fremaux, Hélène Deveau, Melissa Richards, Patrick Boyaval, Sylvain Moineau, Dennis A. Romero, and Philippe Horvath, "CRISPR Provides Acquired Resistance against Viruses in Prokaryotes," *Science* 315, no. 5819 (2007): 1709–12, doi.org/10.1126/science.1138140.

208 **Among the invertebrates, the greatest:** Konstantina T. Tsoumani, Angela Meccariello, Kostas D. Mathiopoulos, and Philippos Aris Papathanos, "Developing CRISPR-Based Sex-Ratio Distorters for the Genetic Control of Fruit Fly Pests: A How To Manual," *Archives of Insect Biochemistry and Physiology* 103, no. 3 (March 2020): e21652, doi.org/10.1002/arch.21652.

208 **One example is removing genes:** Andrew Hammond, Roberto Galizi, Kyros Kyrou, Alekos Simoni, Carla Siniscalchi, Dimitris Katsanos, Matthew Gribble, et al., "A CRISPR-Cas9 Gene Drive System Targeting Female Reproduction in the Malaria Mosquito Vector *Anopheles gambiae*," *Nature Biotechnology* 34 (2016): 78–83, doi.org/10.1038/nbt.3439.

Jieyan Chen, Junjie Luo, Yijin Wang, Adishthi S. Gurav, Ming Li, Omar S. Akbari, and Craig Montell, "Suppression of Female Fertility in *Aedes aegypti* with a CRISPR-Targeted Male-Sterile Mutation," *PNAS* 118, no. 22 (June 2021): e2105075118, doi.org/10.1073/pnas.2105075118.

208 **The next level is to:** Roberto Galizi, Andrew Hammond, Kyros Kyrou, Chrysanthi Taxiarchi, Federica Bernardini, Samantha M. O'Loughlin, Philippos-Aris Papathanos, Tony Nolan, Nikolai Windbichler, and Andrea Crisanti, "A CRISPR-Cas9 Sex-Ratio Distortion System for Genetic Control," *Scientific Reports* 6 (2016): 31139, doi.org/10.1038/srep31139.

208 **Another example of a useful:** Tessema Aynalem, Lifeng Meng, Awraris Getachew, Jiangli Wu, Huimin Yu, Jing Tan, Nannan Li, and

Shufa Xu, "StcU-2 Gene Mutation via CRISPR/Cas9 Leads to Misregulation of Spore-Cyst Formation in *Ascosphaera apis*," *Microorganisms* 10, no. 10 (October 2022): 2088, doi.org/10.3390/microorganisms 10102088.

209 **For example, the heartbeats:** S. M. Ford, M. Watanabe, and M. W. Jenkins, "A Review of Optical Pacing with Infrared Light," *Journal of Neural Engineering* 15, no. 1 (February 2018): 011001, doi.org/10.1088/1741-2552/aa795f.

Aristides B. Arrenberg, Didier Y. R. Stainier, Herwig Baier, and Jan Huisken, "Optogenetic Control of Cardiac Function," *Science* 330, no. 6006 (November 2010): 971–74, doi.org/10.1126/science.1195929.

Yue Cheng, Haitao Li, Hong Lei, Chan Jiang, Panpan Rao, Long Wang, Fang Zhou, and Xi Wang, "Flexible and Precise Control of Cardiac Rhythm with Blue Light," *Biochemical and Biophysical Research Communications* 514, no. 3 (June 2019): 759–64, doi.org/10.1016/j.bbrc.2019.05.035.

About the Author

Drew Harvell is Professor Emerita of Ecology and Evolutionary Biology at Cornell University. She is the author of *Ocean Outbreak* and *A Sea of Glass* which were, variously, the winner of the National Outdoor Book Award for Natural History Literature, recipient of the Rachel Carson Environmental Literature Award, one of the year's best 'Art Meets Science' books by *Smithsonian Magazine*, Prose Award winner in Biological Sciences from the Association of American Publishers and recipient of the Ecological Society of America Sustainability Science Award.

Harvell has written for the *New York Times*, *Seattle Times* and *CNN*, and her work has been featured in the *Atlantic*, *Guardian*, *Washington Post*, *Scientific American*, *Nature* and more. She also featured in the award-winning film *Fragile Legacy* and is currently a science adviser for Fabian Cousteau's Underwater Space Station.